Natural Fibre Biocomposites

Natural Fibre Biocomposites

Editor

Sheldon Q. Shi

MDPI • Basel • Beijing • Wuhan • Barcelona • Belgrade • Manchester • Tokyo • Cluj • Tianjin

Editor
Sheldon Q. Shi
Mechanical Engineering Department,
University of North Texas
USA

Editorial Office
MDPI
St. Alban-Anlage 66
4052 Basel, Switzerland

This is a reprint of articles from the Special Issue published online in the open access journal *Fibers* (ISSN 2079-6439) (available at: https://www.mdpi.com/journal/fibers/special_issues/Natural_Fibre_Biocomposites).

For citation purposes, cite each article independently as indicated on the article page online and as indicated below:

LastName, A.A.; LastName, B.B.; LastName, C.C. Article Title. *Journal Name* **Year**, *Article Number*, Page Range.

ISBN 978-3-03943-210-3 (Hbk)
ISBN 978-3-03943-211-0 (PDF)

Contents

About the Editor

Sheldon Q. Shi, Professor, Mechanical Engineering Department, University of North Texas. Fellow of Society of Wood Science nd Technology (SWST). Dr. Shi obtained his Ph.D. degree at Michigan Technological University (MTU) in 1997. After about two years of research experience as a Post-Doctorate Research Associate at both MTU and University of Maine, he joined APA—The Engineered Wood Association in 1999 as a Scientist/Engineer conducting product evaluation and quality assurance of engineered wood-based products for building construction. In 2004, he joined the faculty at Mississippi State University (MSU) as a faculty and received tenure. After seven years of service with MSU, Dr. Shi joined the University of North Texas. Dr. Shi has been experienced in the manufacturing processes of composite materials using biomass as a feedstock, such as wood, plant fibers, soybean, etc. His expertise is on the processing and evaluation of bio-based composites, adhesive and adhesion, bond durability of the composite materials, and biomass to carbon conversion. He served on the Executive Board of the Society of Wood Science and Technology (SWST) during 2008–2015, and was voted as the President for the society during 2013–2014. Dr. Shi has published over 200 peer-reviewed papers, and over 100 proceedings and presentations.

Preface to "Natural Fibre Biocomposites"

In recent years, there has been significant progress in biomass research as bio-based products are beneficial to the environment, energy-saving, and cost-saving if they are processed properly. The book collects the most state-of-the-art works related to the natural fiber composites in a Special Issue entitled "Natural Fiber Biocomposites". These works address all the issues related the manufacturing of natural fiber composite products, from (1) raw materials, such as wood, flax fiber, and cellulose nanofiber; to the (2) raw material treatments, such as furfuryl alcohol pretreatment, ultrasonic vibration treatment (UVT), and extraction method for the resins; to the (3) process of the composites fabrication, such as thermo-hygro-mechanical densification; and to the (4) performance of the composites, including mechanical, moisture absorption, opacity, thermal, and biodegradability. Discussions on the adhesives/resins used in the natural fiber composites fabrication, such as dried distiller's grains and solubles (DDGS), pennycress (Thlaspi arvense L.) press cakes (PPC), and lesquerella [Lesquerella fendleri (A. Gary) S. Watson] press cake (LPC), starch, and polylactic acid (PLA), are also part of the book. It is believed the technical information presented in this book will contribute to the development of the bio-based composites.

Sheldon Q. Shi
Editor

fibers MDPI

Article

Effect of Density and Fiber Size on Porosity and Thermal Conductivity of Fiberboard Mats

Pamela Rebolledo [1], Alain Cloutier [1,*] and Martin-Claude Yemele [2]

1 Centre de Recherche sur les Matériaux Renouvelables (CRMR), Université Laval, 2425 rue de la Terrasse, Québec, QC G1V 0A6, Canada; pamela.rebolledo-valenzuela.1@ulaval.ca

2 Direction de la Modernisation de L'industrie des Produits Forestiers, Ministère des Forêts, de la Faune et des Parcs, 5700, 4e avenue Ouest, bureau A-202, Québec, QC G1H 6R1, Canada; martin-claude.yemele@mffp.gouv.qc.ca

* Correspondence: alain.cloutier@sbf.ulaval.ca; Tel.: +1 418 656 5851

Received: 31 August 2018; Accepted: 14 October 2018; Published: 19 October 2018

Abstract: The thermal conductivity and porosity of fiberboard mats are crucial parameters for efficient energy consumption of the hot-pressing process and for final panel quality. In this work, the effect of fiber size and mat density on porosity and thermal conductivity of the mat were investigated. The fiber size was characterized as fine, medium and coarse. The mat porosity was measured by image analysis using the black and white contrast method. The thermal conductivity was determined at different density levels with a temperature gradient of 1.6 °C mm^{-1} and 7.6% (s = 0.3) moisture content. The results showed that fiber size was a dominant variable governing heat conduction and mat porosity. The mats made with medium size fibers showed a higher resistance to compression. The thermal conductivity of coarse fiber mats decreased drastically between 700 kg m^{-3} and 810 kg m^{-3}. This was likely due to a higher fracture frequency observed for coarse fibers in comparison to the other fiber sizes studied. Hence, the fine and medium fibers conducted heat more efficiently. Moreover, fiber bundles and fractured fibers were observed during the mat porosity measurements, principally in mats made with fine fiber size.

Keywords: mat porosity; mat thermal conductivity; fiber size; density; hot-pressing process

1. Introduction

Medium density fiberboard (MDF) is a wood-based composite widely used in residential applications and recognized for its easy machining, excellent surface properties and finishing quality. In the industrial manufacture of MDF, hot pressing is a key step due to its impact on the performance of the final panel and the plant productivity. Hot pressing is a complex process because the thickness, density and internal structure of the fiberboard mat change during compression. There is a common agreement that hot pressing of fiber mats involves two different mechanisms: Heat and mass transfer and mat rheological behavior [1–5]. During hot pressing, heat and mass transfer take place in the fibrous material and in the void spaces of the fiber mat. This heat is transferred from press platens to the fiber mat by conduction. The heated fibers generate vapor due to the evaporation of the moisture contained in the fibers. The vapor migrates through void spaces due to a differential vapor pressure between the fibers and the void spaces. Then, part of this vapor is condensed and absorbed partially by the fibers and the remaining vapor flows through the void network by convection and migrates to the environment [1,6,7]. Therefore, mat porosity may affect this moisture transfer and the mechanical behavior of the fiber mat during the development of the density profile through the mat thickness.

The characterization of the mat thermal conductivity and porosity is fundamental because these mat properties are needed in hot-pressing simulation models. An important issue for heat and mass transfer is the behavior of the mat internal structure during hot pressing. The dominant mechanisms of heat and mass transfer during hot pressing are convection of water vapor and heat conduction [1,4,8]. Thermal conductivity determines the conductive heat transfer from the press platens to the mat core and the edges of the fiber mat [8,9]. Besides, convective heat and vapor transfer take place mainly within the void network in the fiber mat [10].

Thermal conductivity is a physical property characterizing the ability of a material to conduct heat. Siau [11] has described thermal conductivity as a relationship between heat flux and the temperature gradient. Later, Von Haas [1] has defined the thermal conductivity of MDF as a function of density as follows:

$$k_t^{0.30} = 4.86 \times 10^{-8}\rho^2 + 4.63 \times 10^{-5}\rho + 4.38 \times 10^{-2} \quad (1)$$

where ρ is mat density (kg m^{-3}).

Given that during hot pressing, heat flows by conduction in the fiber mat, fiber size may have a significant impact. The characterization of the effect of particle size on heat conduction through mat thickness is essential to improving the energy efficiency of hot pressing. Previous works performed by Kavazović et al. [9] and Hata et al. [12] have shown that particle size could affect the physical mechanisms involved during hot pressing and also the quality of simulation models' predictions.

Mat porosity is related to transfer mechanisms and to the density profile development. A closer look at the effect of fiber size and density on mat porosity could help to determine the optimal conditions for the vertical density profile development, improving the final properties of fiberboard. However, few studies have been devoted to this topic. Kamke and Zylkowski [13] have proposed that the void network of the fiberboard mat is more homogeneous than other wood-based composite materials such as oriented strandboard (OSB) and particleboard mats because of a more homogeneous particle size. Delisée et al. [14] have proposed that during the first stages of mat compression, the reduction of the void space between fibers could explain the reduced volume of the mat. Inside this network, fiber bundles produce irregular structures in the mat. Moreover, in fiberboard manufacturing, long fibers are often desirable because they favor a better contact between fibers. This allows increasing the heat conduction during hot pressing. Nevertheless, small fibers may have an impact on the structure of the fiberboard mat, improving its compression efficiency [15,16].

Some studies have suggested that the fibrous structure of the panel affects its thermal conductivity. For instance, Kamke and Zylkowski [13] have found that at the same density, wood-based panels have a lower thermal conductivity than solid wood most likely due to the more continuous structure of solid wood. Suleiman et al. [17] proposed that the dominant factors for thermal conductivity of the mat could be mat porosity and thermal conductivity of the particles. Belley [18] suggested that mat density and fiber size effects may interact with the thermal conductivity of the mat.

The objective of this study was to evaluate the effect of fiber size and mat density on porosity and thermal conductivity of the mat. The relationship between mat porosity and heat conduction was also investigated. Accordingly, this study contributes to characterizing the relationship between various levels of mat compression and fiber size on the evolution of controlling properties of the temperature and internal gas pressure of the mat during hot pressing. This work is included in a broader research project aimed at characterizing the heat and mass transfer phenomena involved in the hot pressing process of fiberboard for use in a simulation model.

2. Materials and Methods

White spruce (*Picea glauca* (Moench) Voss) fibers obtained from wood chips at green-state of trees harvested at the Petawawa Research Forest, Mattawa, Ontario, Canada were used for fiberboard manufacturing.

2.1. Refining and Characterization of Fiber Size

Fiber refining was performed at FPInnovations, Québec City, Québec, Canada. For this process, an Andritz disk refiner with a disk diameter of 55.9 cm was used. Thus, 200 kg of air-dried white spruce chips were used to produce 150 kg of refined fibers. For panel manufacturing, an average of 2 to 3 kg of fibers was used for each panel. Accordingly, we used a total of 135 kg of refined fibers. The same refining temperature and pressure were applied to produce different fiber sizes, but the spacing and speed of the refiner disks were changed as presented in Table 1 [19]. The refined fibers were shipped to the Renewable Materials Research Center, Laval University, Québec, Canada.

Table 1. Refining parameters for white spruce chips.

Fiber Size	Plates Rotation Speed (rpm)	Pressure (bars)	Holding Time (s)	Plate Spacing (mm)	Temperature (°C)
Fine	2750	8	120	0.10	170
Medium	2000	8	120	0.25	170
Coarse	2000	8	120	0.60	170

The fiber sizes were characterized. The fibers were sieved using a US Electrical Motors sieving machine model F844-SMA with a set of seven mesh openings (Fisher Scientific Company, Ottawa, ON, Canada), different for each fiber size according to Figure 1. Samples of 0.500 kg of each fiber size were sieved. Mesh openings between 0.044 mm and 3.00 mm were used [19]. During the fiber characterization, the screens were ordered from smallest to largest. A higher proportion of the total fiber weight was obtained in sieves with a large mesh opening for the three fiber sizes. However, for coarse fibers, a higher proportion of the total fiber weight was noticed in the smallest sieve (fiber size lower than 0.710 mm). For medium fibers, the highest percentage of weight was observed for mesh openings between 0.60 and 0.71 mm. For the fine fibers, the higher fiber weight was obtained for mesh openings between 0.13 and 0.14 mm. Therefore, a fiber size of 0.13 mm for fine fiber size, 0.71 mm for medium fiber size and 3.00 mm for coarse fiber size was considered to manufacture the fiberboards used in this work.

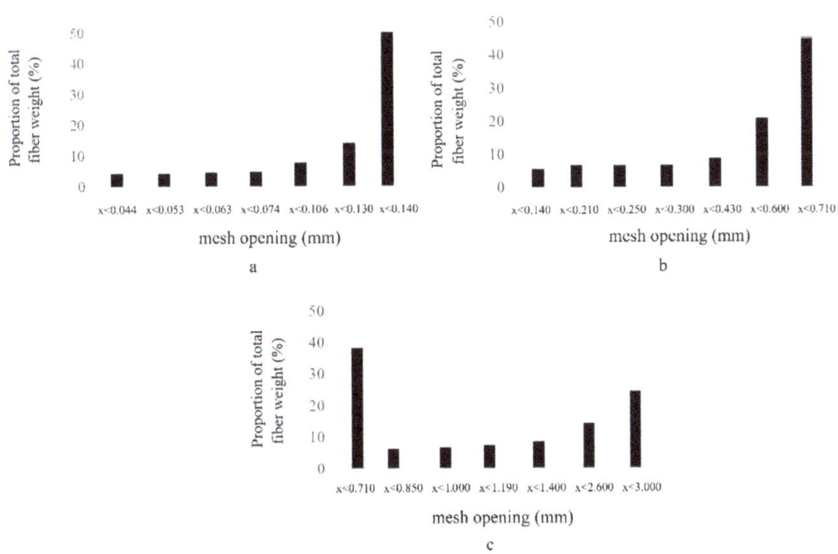

Figure 1. Fiber size distribution of white spruce according to the mesh opening [19]: (**a**) Fine fiber size, (**b**) Medium fiber size, (**c**) Coarse fiber size.

2.2. Homogeneous Density Profile Approach and Density Profile Measurements

In the present work, laboratory panels with homogeneous or uniform density profile through their thickness were manufactured. This method allows the removal of the effect of density variation through the panel thickness. Hence, a total of 45 fiberboards were manufactured with this method at three different fiber sizes and five density levels: 200 kg m^{-3}, 400 kg m^{-3}, 600 kg m^{-3}, 800 kg m^{-3} and 1000 kg m^{-3}. Thus, each density level corresponded to a specific level of mat compression depicting the mat local density produced during hot pressing. The laboratory panels were prepared as follows [19]: First, the fibers were dried to a final moisture content of 2 to 3% in a lumber dryer with a capacity of 2.5 m^3 at a temperature of 60 °C and a relative humidity of 17%. The dried fibers were stored in plastic bags to prevent any moisture uptake. Dried fibers were blended with 14% formaldehyde urea adhesive (UF) based on wood oven-dry mass at 70% solid content, 1% wax based on wood oven-dry mass at 58% solid content, 14% distilled water based on oven-dry mass and 0.2% liquid UF adhesive basis of ammonium chloride at a concentration of 25% to reduce the pH to 5 [19]. The equipment used for blending was an in-house built drum blender equipped with pneumatic atomizers. After blending, the fibers tended to form small balls. They were then passed through a screen refiner (Model PSKM6-350, Pallman, Zweibrücken, Germany) to eliminate these balls. After this process, the fine and medium size fibers had maintained their size distribution. However, the coarse fibers showed a more homogeneous size distribution through the different mesh openings. Next, 560 mm × 460 mm mats were hand-formed in a mold and pre-pressed to a thickness of 120 mm at room temperature using an in-house built cold. The pre-pressing was performed to reduce the closure time in the hot-pressing process. A hot press (Dieffenbacher, Alpharetta, GA, USA) temperature probe was located at the center of the mat to monitor core temperature and vapor pressure (Figure 2). The pre-pressed mats were then hot-pressed with the press platens temperature increasing from 86 °C to 103 °C to obtain the flat vertical density profiles desired. The mats were pressed to a target thickness of 16 mm in about 75 min. Once the target temperature of 86 °C was reached in the core of the mat, the temperature was increased by 5 °C steps every 15 min until 103 °C was reached in the core of the mat because the polymerization temperature of urea formaldehyde is about 100 °C (with NH_4Cl catalyst). The panel was then kept in the hot press at 103 °C for 3 min [19]. The total manufacturing duration for each panel was 1 h approximately. Next, the laboratory panels were conditioned for 14 days to 65% relative humidity and 21 °C temperature.

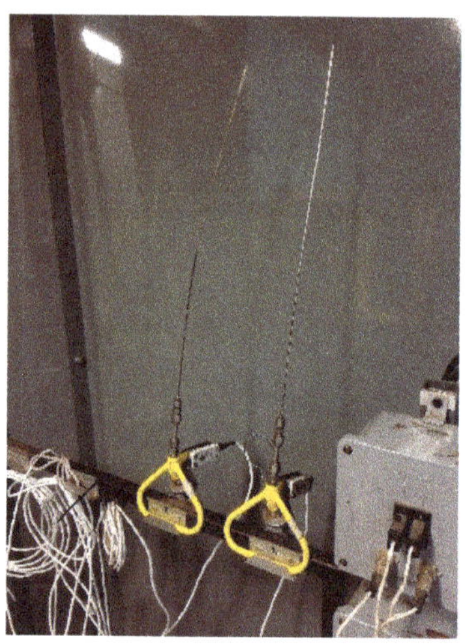

Figure 2. Probes used for mat temperature measurements.

The density profile was determined for each specimen extracted from each laboratory panel. Six specimens of 16 mm thickness and 50 mm diameter were extracted from each panel and conditioned to 65% relative humidity and 21 °C temperature. An X-ray densitometer (Quintek Measurement Systems Inc., model QDP-01X, Knoxville, TN, USA) was used to measure the vertical density profile of all the specimens. These measurements were made at intervals of 0.02 mm through the specimen thickness. The density profiles obtained revealed a homogeneous density profile through the mat thickness as expected [19].

2.3. Mat Anhydrous Density Determination

The mat density is a function of oven-dry density and moisture content [5,7]. In the present investigation, the oven-dry or anhydrous density was obtained from Equation (2):

$$\rho_{OD} = \frac{\rho_{mut}}{(1+M)} \tag{2}$$

where ρ_{mat} is mat density (kg m^{-3}), and M is mat moisture content (dimensionless). The anhydrous density was determined for fiber mats made with fine, medium and coarse fiber sizes. A second-degree polynomial equation for thermal conductivity as a function of the anhydrous density was determined for fine, medium and coarse fiber mats.

2.4. Specimen Preparation for Porosity and Thermal Conductivity Determination

In the current study, samples for thermal conductivity and porosity determination were sawn from each panel as presented in Figure 3. For porosity determination, cubes of 1–2 mm^3 were extracted from each laboratory panel of densities 198 kg m^{-3}, 415 kg m^{-3}, 598 kg m^{-3}, 810 kg m^{-3} and 987 kg m^{-3} and fine, medium and coarse fiber sizes (Figure 4a). Eighteen observations were obtained from each treatment (15 treatments in total). This gave a total of 270 cubes for the whole experiment. The cubes were dried to anhydrous conditions and impregnated with LR White medium grade, acrylic resin, model 14381, manufactured by Electron Microscopy Sciences (Hatfield, PA, USA). (Figure 4b). The following procedure was developed in our laboratory. The dried cubes were submerged in acrylic resin. After one hour, the resin was purged. This treatment was applied three times. Subsequently, the cubes were submerged again in the acrylic resin for 24 h. The resin was purged and each cube was placed inside a clear gelatine capsule, 2 oz. size, model 7, manufactured by Galenova Inc (St-Hyacinthe, QC, Canada) and filled with resin. Then, each capsule was covered with a gel cover and immediately placed inside an oven at 60 °C for 24 h to polymerize the resin. Later, these capsules were submerged in cold water for 10 min for gel softening. Next, the gel was retired entirely from the resin capsule surface already hardened (Figure 4c). Afterward, three thin slices of 3 µm in thickness were cut with a microtome from each impregnated cube. A total of 810 thin slices were obtained for porosity measurement. Each slice was stained with a solution of methylene-blue at 1%, blue-methyl azur II at 1% and water at neutral pH. Next, each slice was put on a microscope slide forming a microtomed sample. The final dimension of each microtomed specimen was about 1.44 mm × 1.83 mm × 3 µm (Figure 4d).

Specimens of 152.4 mm × 152.4 mm × 16 mm for thermal conductivity determination were conditioned to 65% relative humidity and 21 °C temperature in 21 days. Average moisture content of these specimens was 7.6% (s = 0.3) which was determined by anhydrous mass method. Twelve observations were considered for each of the 15 treatments, which gave a total of 180 observations in total for this test.

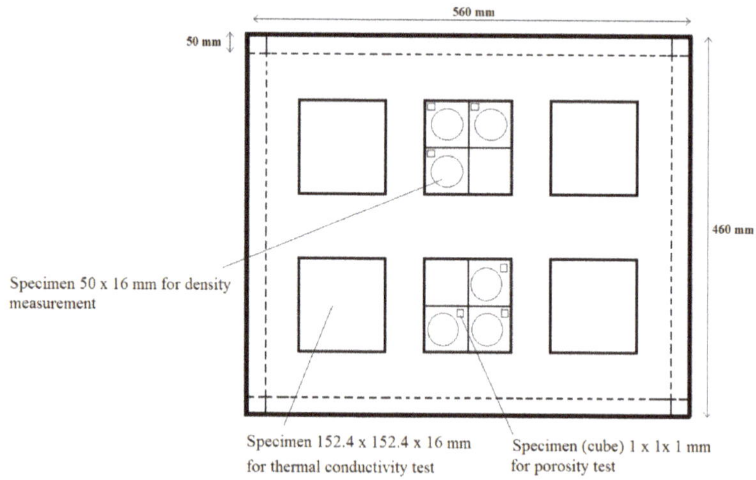

Figure 3. Specimen cutting scheme in each laboratory panel.

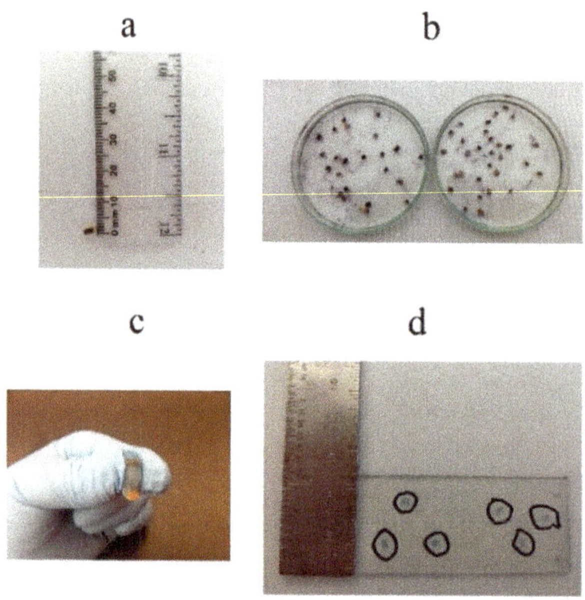

Figure 4. Impregnation procedure of the fiberboard specimens for porosity determination: (**a**) Cutting a fiberboard cube of 1 mm^3, (**b**) Submersion of fiberboard specimen in acrylic resin, (**c**) Fiberboard specimen in polymerized resin, (**d**) Microtomed slices obtained from impregnated specimens.

2.5. Porosity Determination

The porosity determination was performed using an image analysis method. This procedure consisted in measuring void spaces on an image taken from each microtomed sample (Figure 4d) using black and white contrast. Each image displayed on the screen was divided into four zones [20] during optical microscope observation (Nikon, model eclipse E600, Mississauga, ON, Canada) with a magnification of 40×. This equipment is presented in Figure 5. One image was taken for each of these zones with a Pixelink, model PL-A686C camera (Ottawa, ON, Canada), with a resolution of 2208 × 3000 pixels and an image size of 7.7 mm × 10.5 mm. Due to some noise and impurities, the images were corrected and improved with the Adobe Photoshop program, version CS6 Extended (Softchoice, Québec, QC, Canada). For each microtomed sample, four images were obtained, which gave a total of 3240 images. The porosity measurements were made for each of these images. Four images were taken for each microtomed sample according to Figure 6. The average value of these four images was considered as the porosity of the sample. When all the images were obtained, the porosity was calculated by black and white contrast with the Wincell program, version v2004a, manufactured by Regent Instruments Inc. (Softchoice, Québec, QC, Canada). This software calculated the percentage of black pixels in the image.

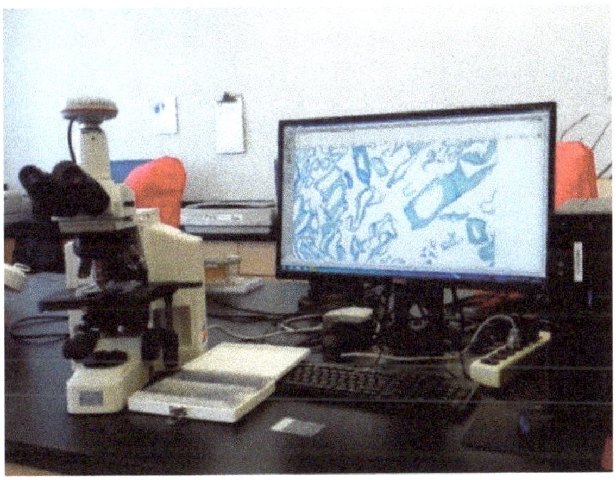

Figure 5. Setup used to measure mat porosity by image analysis: Optical microscope, camera, and Wincell software.

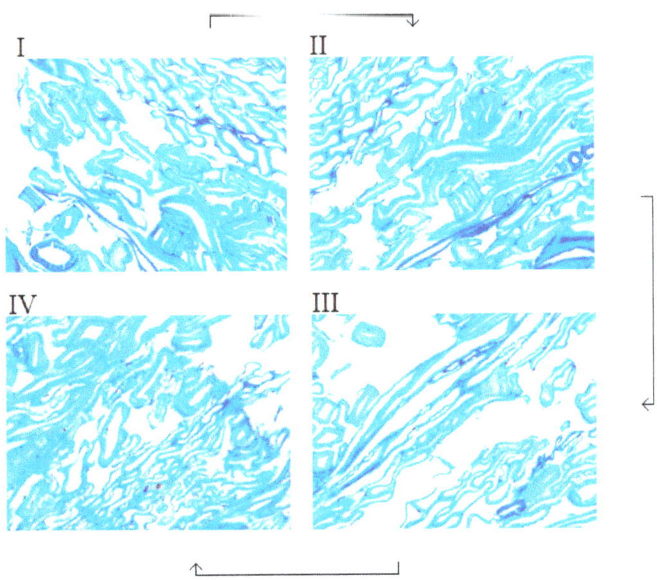

Figure 6. Four images were taken in one fiberboard slice for analyses with the Wincell software. The arrows indicate the porosity measurement sequence. I: first measurement; II: second measurement; III: third measurement; IV: fourth measurement.

2.6. *Thermal Conductivity Determination*

The thermal conductivity of 180 specimens was measured. The specimens were previously conditioned to 65% relative humidity and 21 °C temperature for 21 days. The measurements of thermal conductivity were determined according to the steady-state thermal resistance procedure decribed in ISO standard [21]. This test was performed with a LaserComp Fox 314 (TA instruments, New Castle, DE, USA) apparatus as shown in Figure 7. Each 16 mm thick specimen of was placed inside of this equipment. The upper and lower plates of the thermal conductivity equipment were heated at 10 °C and 35 °C respectively, corresponding to a temperature gradient of 1.6 °C mm^{-1}. The measurement duration increased when the density level of the specimen decreased. For instance, with a low-density specimen (198 kg m^{-3} and 415 kg m^{-3}) the measurements stabilization took 5 to 6 h compared to 1 h for a higher density specimen.

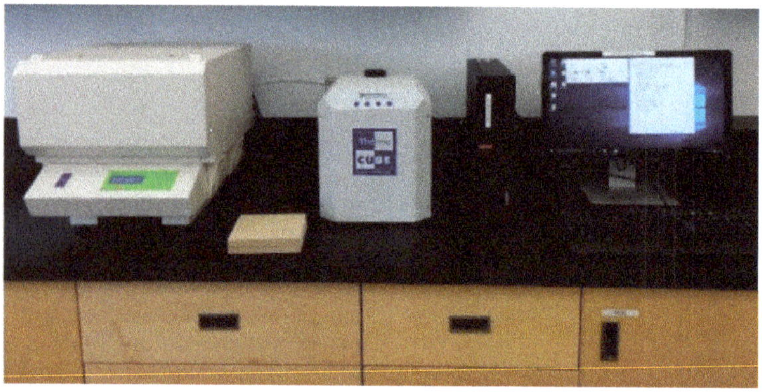

Figure 7. Equipment for thermal conductivity measurements.

2.7. *Experimental Design*

A full factorial design was used. The effects of two sources of variation, namely mat density and fiber size on thermal conductivity and mat porosity were examined. Tests of normality and variance homogeneity were performed to validate the hypotheses of the analysis of variance. Five density levels (198 kg m^{-3}, 415 kg m^{-3}, 598 kg m^{-3}, 810 kg m^{-3} and 987 kg m^{-3}) and three fiber sizes (fine, medium and coarse) were studied for a total of 15 treatments. Each treatment was replicated three times. Additionally, a regression analysis was performed to investigate the porosity—thermal conductivity relationship. Statistical analyses were performed in the SAS 9.4 (SAS Institute Inc., Cary, NC, USA) software. Regression analyses were made in the MATLAB (version R2016a, Natick, MA, USA) numerical analysis software.

3. Results

3.1. *Mat Porosity and Fiber Bundles*

The evolution of porosity and thermal conductivity of the mat during compression was analyzed as a function of mat density. For this, it should be kept in mind that during hot pressing, a vertical density profile through the mat thickness is developed. Therefore, the local density evolution can be linked to local mat thermal conductivity and porosity. Hence, the knowledge of how mat density is related to these mat properties could help to understand how the vertical density profile is formed and improve simulation models' predictions.

Figure 8 shows the black and white images used to calculate mat porosity. The porosity measurements were made with a magnification of 40× as mentioned. The black and white pixels in the images correspond to void spaces and fibrous material respectively.

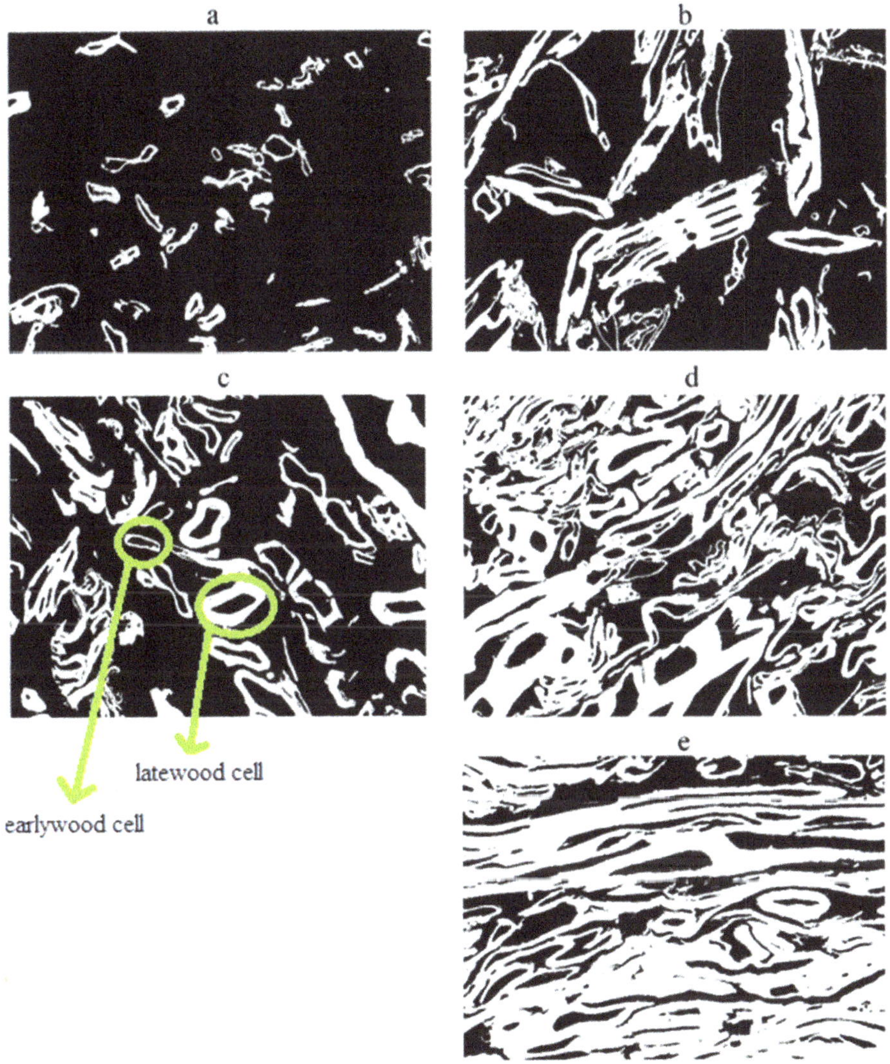

Figure 8. Images (40×) used to calculate mat porosity by the black and white contrast method at different density levels: (**a**) 198 kg m^{-3}, (**b**) 415 kg m^{-3}, (**c**) 598 kg m^{-3}, (**d**) 810 kg m^{-3}, (**e**) 987 kg m^{-3}. White is fibrous material, and black is void spaces.

Most of the porosity images presented fiber bundles or shives, as shown in Figure 9a,b. This was observed at 20× and 40× magnifications.

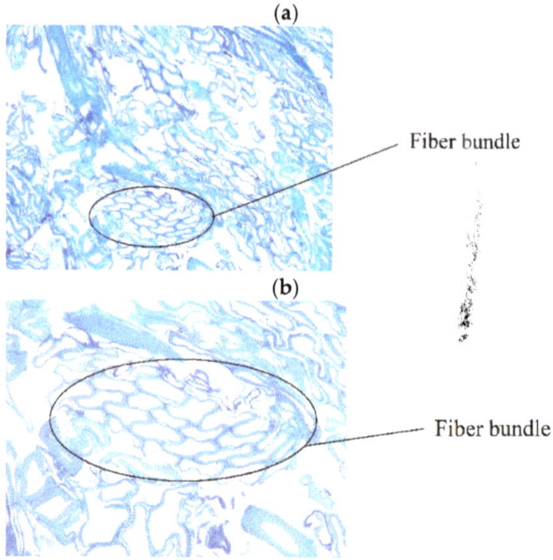

Figure 9. Images of mat porosity at 810 kg m^{-3} with the same fiber bundle at different magnifications: (**a**) 20×, (**b**) 40×.

3.2. *Effect of Fiber Size and Density on Mat Porosity*

Table 2 shows the analysis of variance performed to evaluate the influence of density and fiber size on mat porosity. The *p*-value indicated a significant interaction between the effects of mat density and fiber size (*p*-value < 0.0001) on porosity. Figure 10 shows that the curves of fine, medium and coarse fiber sizes were not parallel. This demonstrates the interaction between variables fiber size and mat porosity, presented in Table 2, which indicates that the effect of density on mat porosity varies with fiber size.

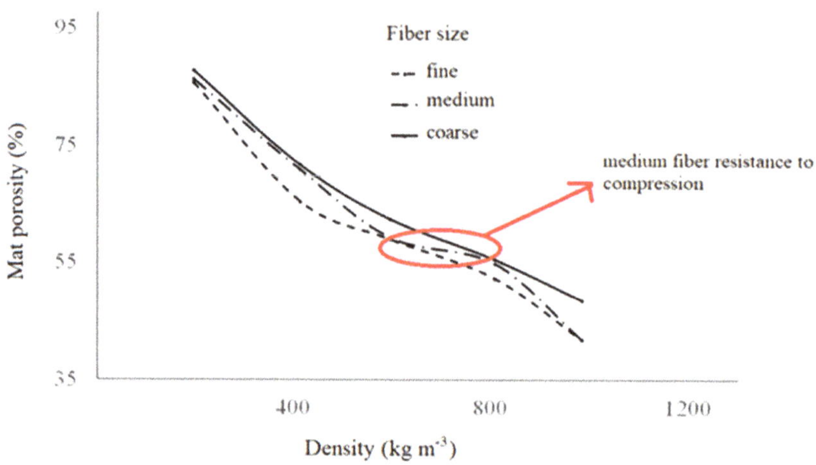

Figure 10. Interaction of mat porosity versus density at different fiber sizes.

Table 2. Analysis of variance performed for the mat porosity at different density levels and fiber sizes.

Source	Degrees of Freedom	Sum of Squares	Mean Square	F-Value	p-Value
Mat density	4	670,202	167,550	3907.4	<0.0001
Fiber size	2	10,660	5330	124.3	<0.0001
Mat density × fiber size	8	4515	564	13.2	<0.0001

As expected, mat porosity decreased from 198 kg m^{-3} to 987 kg m^{-3} (Figure 10). However, the porosity decreased more quickly in mats made with fine fibers. This suggested that the mats made with fine fibers presented a lower resistance to compression. The mats made with medium and coarse fibers presented a higher resistance to compression.

3.3. Effect of Fiber Size and Density on Thermal Conductivity of the Mat

An analysis of variance was also performed to investigate the effect of mat density and fiber size on mat thermal conductivity (Table 3). The results showed a very significant interaction between mat density and fiber size. The interaction between the variables mat density and fiber size is illustrated in Figure 11 by the curves obtained for fine, medium and coarse fibers, which are not parallel. This indicated that heat conduction during mat compression is dependent on mat density and fiber size.

Table 3. Analysis of variance for thermal conductivity at different levels of mat density and fiber sizes.

Source	DF	Sum of Squares	Mean Square	F Value	Pr > F
Mat density	4	0.25554	0.06388	5292.1	<0.0001
Fiber size	2	0.00016	0.00008	6.7	0.0020
Mat density × fiber size	8	0.00041	0.00005	4.28	0.0001

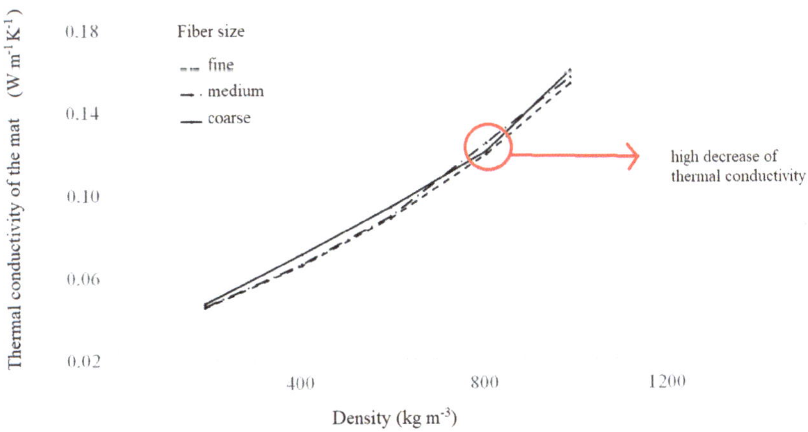

Figure 11. Interaction of thermal conductivity of the mat versus density at different fiber sizes.

Coarse fibers thermal conductivity is higher than for medium and fine fibers according to Figure 11. At higher mat densities up to 810 kg m^{-3}, thermal conductivity continued to increase but more slowly for coarse fiber mats. A low difference of heat conduction between fiber size through the mat thickness was observed at low densities (around 200 kg m^{-3}).

Figure 12 shows the thermal conductivity obtained for fine, medium and coarse fiber sizes against anhydrous mat density calculated from Equation (2). In the fine fiber size group, the highest variability of heat conduction was observed around 600 kg m^{-3}. The variability in medium fiber size group increased with anhydrous density as well. In the case of the coarse fiber group, a higher variability can be noticed at 1000 kg m^{-3}.

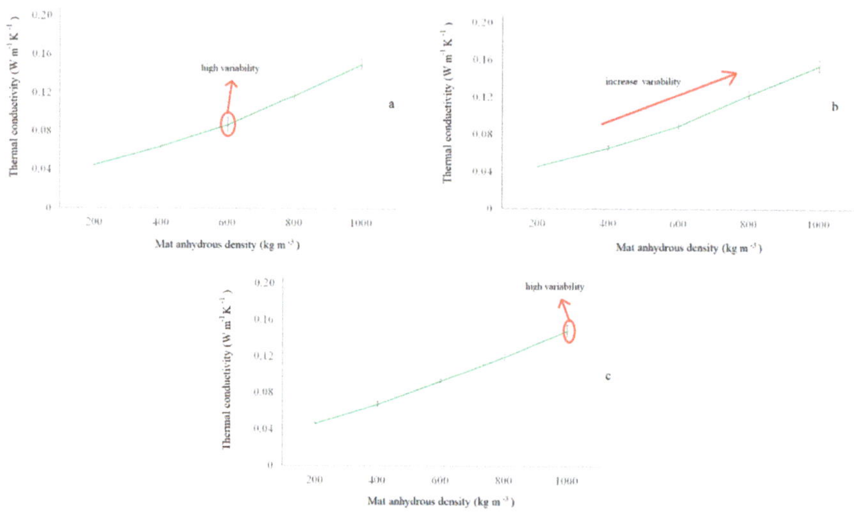

Figure 12. Thermal conductivity versus mat anhydrous density for different fiber sizes: (**a**) Fine size, (**b**) medium size, (**c**) coarse size. Vertical lines show error bars.

A second-degree polynomial expression was fitted to the experimental data of thermal conductivity as a function of oven-dry density for each fiber size studied (Equation (3)).

$$TC = \alpha_1 \rho_{OD}{}^2 + \alpha_2 \rho_{OD} + \alpha_3 \tag{3}$$

where TC is thermal conductivity of the fiber mat (W m^{-1}K^{-1}), ρ_{OD} is anhydrous density of the fiber mat (kg m^{-3}) and α_1, α_2 and α_3 are coefficients depending on the fiber size. These coefficients are presented in Table 4 for the three fiber sizes considered. The coefficient of determination R^2 of the three expressions was 0.98–0.99, which indicated that only 1–2% of the total thermal conductivity variation observed could not be explained by these models.

Table 4. Regression equation coefficients to calculate the thermal conductivity of the fiber mat as a function of mat anhydrous density (Equation (3)).

Fiber Size	α_1	α_2	α_3	R^2
Fine	8.13×10^{-8}	5.28×10^{-5}	0.03	0.98
Medium	7.81×10^{-8}	6.22×10^{-5}	0.03	0.99
Coarse	6.27×10^{-8}	7.25×10^{-5}	0.03	0.99

3.4. Thermal Conductivity in Relation to Mat Porosity

Figure 13 presents the thermal conductivity against mat porosity. A second-degree polynomial equation was fitted to the thermal conductivity porosity experimental data. The determination coefficient R^2 obtained was 0.97, which indicated that 97% of the total variation of thermal conductivity observed was explained by this model.

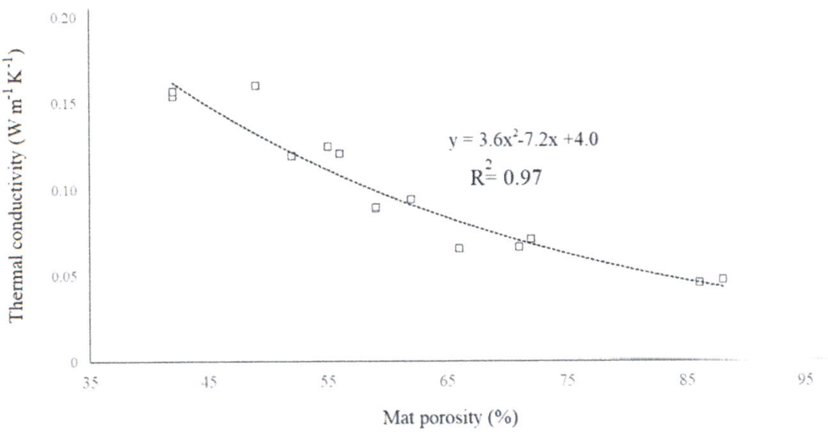

Figure 13. Effect of porosity on the thermal conductivity of the mat for all fiber sizes.

4. Discussion

Figure 9 presents images used in the mat porosity measurements by the black and white contrast method. Collapse and fracture in latewood, as well as in earlywood cells were observed, principally between 810 kg m^{-3} and 987 kg m^{-3} of density. However, latewood cells have preserved their shape much more than earlywood cells. This higher deformation in earlywood cells is most likely due to its thinner cell wall thickness proposed by Dogu et al. [22] and Budakci et al. [23]. Furthermore, a progressive reduction of the void spaces between fibers from 198 kg m^{-3} to 987 kg m^{-3} was observed, as well as the tendency of fibers to preserve their shapes. These findings are in agreement with works performed by Delisée et al. [14] and Tran et al. [24] on low-density insulation panels. When the densification level of fiber mats increased, the fibers' orientation in one direction also seemed to increased (Figure 7e). A notable change from a random orientation towards a better orientation of fibers also seems to occur from 598 kg m^{-3} to 987 kg m^{-3}. Pettersson et al. [25], Zhou et al. [26], Dogu et al. [22] have suggested a similar behavior when the oriented strand board (OSB) mat is compressed, and the strands begin to align themselves in the same direction. The images at 810 kg m^{-3} and 987 kg m^{-3} (Figure 7d,e) exhibited a more homogeneous and oriented structure in comparison to lower densities.

According to Thoemen et al. [27] and Sliseris et al. [28], the presence of fiber bundles in the mat produces void spaces in its surrounding areas, which could affect the mat's response to compression. Although the latewood and earlywood cells were scattered randomly inside the mat, the inhomogeneities found in the images were due principally to fiber bundles. This could have a significant effect on the physical properties of the mat during compression. A considerable deformation of fiber bundle cells compared to the surrounding cells was also observed (Figure 9b). According to Delisée et al. [14], these differences in the fibrous network could influence the mat response to compression due to the highest level of fiber orientation of bundles in comparison to the individual fibers.

Figure 10 showed the porosity results in relation to mat density. These results are in agreement with findings reported by Siau [11] and Belley [18]. Regarding the effect of mat density and fiber size on mat porosity, the results revealed a significant combined effect of density and fiber size (Table 2) as proposed by Bolton and Humphrey [29] and Dai et al. [30]. The mat porosity decreased significantly when the mat density increased. These results are in agreement with studies performed on wood-based composite materials by Dai et al. [31], Belley [18], Li et al. [32] and Ai et al. [33]. The impact of mat density on porosity was dependent on fiber size (Figure 10). Additionally, it was observed that when the mat density increased from 198 kg m^{-3} to 415 kg m^{-3} the porosity decreased faster in mats made with fine fibers in comparison with medium and coarse fiber mats. Then, when the mat density increased towards 598 kg m^{-3}, the porosity of mats made with medium fibers decreased faster than the mats made with fine and coarse fibers. Next, at higher densities, the porosity decreased in a similar manner for mats made with fine and coarse fibers. The porosity of the mats made with medium fibers decreased more slowly when the density increased towards 810 kg m^{-3} and later it decreased quickly as the mat density reached 987 kg m^{-3}. This showed that the medium fiber mats offered a higher resistance to compression when the mat was compressed from 598 kg m^{-3} to 810 kg m^{-3}. Furthermore, the porosity decreases differently in mats made with fine, medium and coarse fibers due to the high impact of fiber size on the porous structure of the mat. Previous works performed by Bolton and Humphrey [29], Thoemen and Klueppel [15], Sliseris et al. [28] have suggested a particle size effect on the mat structure. This is in agreement with the current results which indicated that fiber size had an impact on void spaces. The fine fibers could tend to agglomerate more easily forming fiber bundles, thus decreasing the void spaces between fibers.

Fiber size and mat density are fundamental variables for the thermal conductivity of the mat because heat conduction is carried out principally through fibers as mentioned above. The results of the analysis of variance (Table 3) showed a high interaction between density and fiber size on heat conduction through mat thickness. Figure 11 presents this interaction. Thermal conductivity observed in coarse fiber mats was significantly higher than in mats made with fine and medium fibers between 198 kg m^{-3} and 598 kg m^{-3}. Nonetheless, between 598 kg m^{-3} and 987 kg $m^{-3,}$ the thermal conductivity of coarse fiber mats decreased unexpectedly. This was probably due to the fracture of the fibers' cell wall as mentioned above (see Figure 8d,e), when the mat was compressed to about 810 kg m^{-3} and its structure became more discontinuous for heat conduction. These fiber fractures could cause a significant change of the fibrous network structure when the mat density changes from 598 kg m^{-3} to around 810 kg m^{-3} as proposed by Bolton and Humphrey [29], Sun et al. [10]. Hence, the mat structure could be more affected in the case of coarse fibers because of their tendency to break, impacting more strongly heat conduction through mat thickness.

Figure 12a–c presented the thermal conductivity of mats made with fine, medium and coarse fiber against anhydrous density respectively. As already mentioned, the variability pattern of the three fiber sizes considered in this investigation was different. The thermal conductivity variability in fine and coarse fiber sizes was higher at 600 kg m^{-3} and 1000 kg m^{-3} respectively, while the variability of thermal conductivity in medium fiber sizes increased with density overall. This also suggested that fiber size had a significant impact on the internal structure of the mat. The quadratic parameter of the regression equations indicates the grade of the parabola curvature. A larger magnitude shows a higher increment of thermal conductivity with a sustained increase in density. The quadratic coefficient corresponding to thermal conductivity in mats made with fine fibers was higher than for mats made with medium and coarse fiber sizes. This indicated that the fine fibers conducted heat more efficiently than medium and coarse fibers most likely due to the lower void spaces. Furthermore, a higher frequency of cell wall rupture of coarse fibers in comparison to fine and medium size fibers could have caused a reduction in the heat conduction capacity of coarse fiber mats, as mentioned above. Additionally, it was found that thermal conductivity increased with an increase of mat densification levels as proposed by Kamke and Zylkowski [13], Zambori et al. [8], Belley [18], Sonderegger and Niemz [34], Suleiman et al. [17], Haselein [35] and Thoemen and Humphrey [2] (Equation (1)) in studies performed on wood and wood-based composites.

As it is known, composites materials such as MDF can be considered as materials with discontinuities in their fibrous structure. Thereby, the thermal conductivity of the fiberboard mat may also be affected by a discontinuous structure. Figure 13 presents thermal conductivity against mat porosity for all fiber sizes. The mat porosity had a substantial effect on thermal conductivity. This tendency was also reported by Dai et al. [31], Thoemen et al. [26], Sonderegger and Niemz [34] and Suleiman et al. [14] in investigations on wood-based composite materials. The drastic decrease of thermal conductivity below 65% of mat porosity was evident. Meanwhile, at mat porosities above this percentage, the thermal conductivity reduction was asymptotic. The heat conduction capacity of fiber mats decreased as mat porosity increased as suggested by Kamke and Casey [36]. When mat porosity increased towards 62%, the thermal conductivity decreased rapidly. This indicated that heat conduction was more effective at low porosity levels when the mat is more compressed, and the void spaces are lower. This could also be due to the collapse and fracture of the fibers below a porosity of 62%, producing a drastic decrease of the void spaces as observed in this study. Conversely, thermal conductivity decreased more slowly when the mat porosity increased above 62%. This indicated that the high proportion of void spaces in the fiber mat had a high impact on heat conduction. Thus, a high proportion of the void spaces between fibers could offer higher resistance to heat conduction than the intra-fibers void spaces, which usually predominate at lower porosity levels [14,24].

5. Conclusions

1. Fiber size affected the porosity and thermal conductivity of the mat. This indicates that fiber size should be considered in predictive models.
2. The fiber mats made with medium fiber size (between 0.60 and 0.71 mm) showed a higher resistance to compression. The mats made with coarse fibers (3.00 mm) presented a higher level of fracture and collapse which affected their capacity for heat conduction.
3. The results of this investigation suggested that heat conduction is more efficient during the compression of mats with fine fibers (0.044–0.14 mm). This indicated that fine fibers had a lower tendency to produce void spaces in the fiber mat.
4. The void space affected heat conduction efficiency through the mat thickness.
5. During the porosity measurements, fiber bundles and fiber cell wall fractures were observed. This had an impact on mat compression and heat conduction of the fiber mats.

Author Contributions: All authors contributed extensively to the presented work, discussed, reviewed, and agreed to the modifications at all stages of the paper. P.R. carried out the experiment and prepared the first version of manuscript. P.R. and A.C. analyzed and discussed the results. A.C. and M.-C.Y. reviewed and corrected the manuscript.

Funding: This research was supported by the Natural Sciences and Engineering Research Council of Canada (NSERC) under Discovery Grant #121954-2012.

Acknowledgments: The authors are grateful to Richard Janvier and David Lagueux for their technical assistance in the samples preparation and porosity measurements respectively.

Conflicts of Interest: The authors declare no conflict of interest.

References

1. Von Haas, G. Investigation of the Hot Pressing of Wood Composite-Mats under Special Consideration of the Compression Behavior, the Permeability, the Temperature Conductivity and Sorption-Speed. Ph.D. Thesis, Hamburg University, Hamburg, Germany, 1998. (In Germany)
2. Thoemen, H.; Humphrey, P.E. Modeling the physical process relevant during hot pressing of wood-based composites. Part I. Heat and mass transfer. *Holz Roh Werkst.* **2006**, *64*, 1–10. [CrossRef]
3. Thoemen, H.; Haselein, C.; Humphrey, P.E. Modeling the physical process relevant during hot pressing of wood-based composites. Part II. Rheology. *Holz Roh Werkst.* **2006**, *64*, 125–133. [CrossRef]
4. Kavazović, Z.; Deteix, J.; Cloutier, A.; Fortin, A. Numerical modeling of the medium-density fiberboard hot pressing process, Part I: Coupled heat and mass transfer model. *Wood Fiber Sci.* **2012**, *42*, 168–188.
5. Kavazović, Z.; Deteix, J.; Fortin, A.; Cloutier, A. Numerical modeling of the medium-density fiberboard hot pressing process, Part II: Mechanical and heat and mass transfer models. *Wood Fiber Sci.* **2012**, *44*, 243–262.
6. Humphrey, P.E. Physical Aspects of Wood Particleboard Manufacture. Ph.D. Thesis, University of Wales, Cardiff, UK, 1982.

7. Kayihan, F.; Johnson, J.A. Heat and moisture movement in wood composite materials during the pressing operation—A simplified model. In *Book Numerical Methods in Heat Transfer*, 2nd ed.; Lewis, K., Morgan, B.A., Eds.; John Wiley and Sons: New York, NY, USA, 1983; Volume 2, pp. 511–531.
8. Zambori, B.G.; Kamke, F.A.; Watson, L.T. Sensibility analysis of internal mat environment during hot pressing. *Wood Fiber Sci.* **2004**, *36*, 195–209.
9. Kavazović, Z.; Deteix, J.; Cloutier, A.; Fortin, A. Sensitivity study of a numerical model of heat and mass transfer involved during the medium-density fiberboard hot pressing process. *Wood Fiber Sci.* **2010**, *42*, 130–149.
10. Sun, G.; Ibach, R.; Faillace, M.; Gnatowski, M.; Glaeser, J.; Haight, J. Laboratory and exterior decay of wood-plastic composite boards: Voids analysis and computed tomography. *Wood Mater. Sci. Eng.* **2017**, *12*, 263–278. [CrossRef]
11. Siau, J. *Wood: Influence of Moisture on Physical Properties*; Department of Wood Science and Forest Products, Virginia polytechnic Institute and State University: Blacksburg, VA, USA, 1995; p. 227.
12. Hata, T.; Kawai, S.; Ebihara, T.; Sasaki, H. Production of particleboards with a steam injection press V. Effects of particle geometry on temperature behaviors in particle mats and on-air permeabilities of boards. *Mokuzai Gakkaishi* **1993**, *39*, 161–168.
13. Kamke, F.A.; Zylkowski, S.C. Effect of wood-based panel characteristics on thermal conductivity. *For. Prod. J.* **1989**, *39*, 19–24.
14. Delisée, C.; Badel, E.; Lux, J.; Malvestio, J. Caractérisation microstructurale 3D et densification locale d'isolants fibreux cellulosiques sollicites en compression. *Eur. J. Environ. Civ. Eng.* **2009**, *13*, 429–442. [CrossRef]
15. Thoemen, H.; Klueppel, A. An investigation on the permeability of different wood furnish materials. *Holzforschung* **2008**, *62*, 215–222. [CrossRef]
16. Sackey, E.K.; Smith, G.D. Characterizing macro-voids of uncompressed mats and finished particleboard panels using response surface methodology and X-ray CT. *Holzforschung* **2010**, *64*, 343–352. [CrossRef]
17. Suleiman, B.M.; Larfeldt, J.; Leckner, B.; Gustavsson, M. Thermal conductivity and diffusivity of wood. *Wood Sci. Technol.* **1999**, *33*, 465–473. [CrossRef]
18. Belley, D. Détermination des Propriétés de Transfert de Chaleur et de Masse des Panneaux de Fibres de bois MDF. Master's Thesis, Univeristé Laval, Québec, QC, Canada, 2009. (In French)
19. Rebolledo, P.; Cloutier, A.; Yemele, M.-C. Gas permeability of fiberboard mats as a function of density and fiber size. *Wood Mater. Sci. Eng.* **2018**, *1*, 1–8. [CrossRef]
20. Chaffey, N. Wood microscopical techniques. In *Wood Formation in Trees*; Chaffey, N., Ed.; Taylors and Francis: New York, NY, USA, 2002; Chapter 3; pp. 17–40.
21. ISO 8302:1991. *Thermal Insulation—Determination of Steady-State Thermal Resistance and Related Properties—Guarded Hot Plate Apparatus*; Technical Committee ISO/TC 163; International Organization for Standardization: Geneva, Switzerland, 2013.
22. Dogu, D.; Tirak, K.; Candan, Z.; Unsal, O. Anatomical investigation of thermally compressed wood panels. *Bioresources* **2010**, *5*, 2640–2663.
23. Budakci, M.; Pelit, H.; Sonmez, A.; Korkmaz, M. The effects of densification and heat post-treatment on hardness and morphological properties of wood materials. *Bioresources* **2016**, *11*, 7822–7838. [CrossRef]
24. Tran, H.; Doumalin, P.; Delisée, C.; Dupre, J.C.; Malvestio, J.; Germaneau, A. 3D mechanical analysis of low-density wood-based fiberboards by X-ray microcomputed tomography and Digital Volume Correlation. *J. Mater. Sci.* **2012**, *48*, 3198–3212. [CrossRef]
25. Pettersson, P.; Staffan Lundström, T.; Wikström, T. A method to measure the permeability of dry fiber mats. *Wood Fiber Sci.* **2006**, *38*, 417–426.
26. Zhou, C.; Dai, C.; Smith, G. A generalized mat consolidation model for wood composites. *Holzforschung* **2008**, *62*, 201–208. [CrossRef]
27. Thoemen, H.; Walther, T.; Wiegmann, A. 3D simulation of macroscopic heat and mass transfer properties from the microstructure of wood fibre networks. *Compos. Sci. Technol.* **2008**, *68*, 608–616. [CrossRef]
28. Sliseris, J.; Handra, H.; Kabel, M.; Dix, B.; Plinke, B. Virtual characterization of MDF fiber network. *Eur. J. Wood Wood Prod.* **2017**, *75*, 397–407. [CrossRef]
29. Bolton, A.J.; Humphrey, P.E. The permeability of wood-based composite materials. *Holzforschung* **1994**, *48*, 95–100. [CrossRef]

30. Dai, C.; Yu, C.; Xu, C.; He, G. Heat and mass transfer in wood composite panels during hot pressing. Part 4. Experimental investigation and model validation. *Holzforschung* **2007**, *61*, 83–88. [CrossRef]
31. Dai, C.; Yu, C.; Zhou, X. Heat and mass transfer in wood composite panels during hotpressing: Part II Modeling void formation and mat permeability. *Wood Fiber Sci.* **2005**, *37*, 242–257.
32. Li, P.; Dai, C.; Wang, S. A simulation of void variation in wood-strand composites during consolidation. *Holzforschung* **2009**, *63*, 357–361. [CrossRef]
33. Ai, W.; Duval, H.; Pierre, F.; Perré, P. A novel device to measure gaseous permeability over a wide range of pressures: Characterization of slip flow for Norway spruce, European beech, and wood-based materials. *Holzforschung* **2017**, *71*, 147–162. [CrossRef]
34. Sonderegger, W.; Nienz, P. Thermal conductivity and water vapour transmission properties of wood-based materials. *Eur. J. Wood Wood Prod.* **2009**, *67*, 313–321. [CrossRef]
35. Haselein, C.R. Numerical Simulation of Pressing Wood-Fiber Composites. Ph.D. Thesis, Oregon State University, Corvallis, OR, USA, 1998.
36. Kamke, F.A.; Casey, L. Fundamentals of flakeboard manufacture: Internal-mat conditions. *For. Prod J.* **1989**, *38*, 38–44.

fibers

MDPI

Article

Influence of Furfuryl Alcohol Fiber Pre-Treatment on the Moisture Absorption and Mechanical Properties of Flax Fiber Composites

Yunlong Jia * **and Bodo Fiedler**

Institute of Polymer and Composites, Hamburg University of Technology (TUHH), Denickestrasse 15, D-21073 Hamburg, Germany; fiedler@tuhh.de

* Correspondence: yunlong.jia@tuhh.de; Tel.: +49-40-42878-2536

Received: 31 July 2018; Accepted: 17 August 2018; Published: 19 August 2018

check for updates

Abstract: Poor moisture resistance of natural fiber reinforced bio-composites is a major concern in structural applications. Many efforts have been devoted to alleviate degradation of bio-composites caused by moisture absorption. Among them, fiber pre-treatment has been proven to be effective. This paper proposes an alternative "green" fiber pretreatment with furfuryl alcohol. Pre-treatments with different parameters were performed and the influence on the mechanical properties of fiber bundles and composites was investigated. Moisture resistance of composites was evaluated by water absorption tests. Mechanical properties of composites with different water contents were analyzed in tensile tests. The results show that furfuryl alcohol pretreatment is a promising method to improve moisture resistance and mechanical properties (e.g., Young's modulus increases up to 18%) of flax fiber composites.

Keywords: durability; green composites; cellulosic fibers; water uptake

1. Introduction

Sustainable development has increasing importance nowadays. Many efforts have been devoted to reducing greenhouse gases emission and avoiding mass production of products from non-renewable resources. In the field of advanced composite materials, bio-composites (we only refer to plant fibers and bio-based polymers in this work) and their applications gain increasing interest in many industries [1–3]. Apart from the self-evident advantage of reduced environmental impact, bio-composites exhibit acceptable specific mechanical properties [4,5]. Plant fibers have other advantages of low cost, low density, easy handling, non-abrasive nature, etc. [6,7]. Their applications in composites can now be found in automotive, sports, transport vehicles and construction industries [8–10].

Despite the promising mechanical properties of bio-composites that have been proven by many researchers, applications of bio-composites in structural components are still hindered by several challenges [6,11]. Major concerns lie in the hydrophilia of plant fibers, which brings two problems: the poor compatibility between hydrophilic plant fibers and hydrophobic polymer matrix and the low moisture resistance of bio-composites, which is related to durability of bio-composites. Indeed, plant fibers, as cellulosic fibers, are abundant in polar functional groups such as hydroxyl groups. When bio-composites are exposed to humid atmosphere or rain, absorbed water molecules in the composites are very susceptible to form hydrogen bonds with these polar functional groups and then remain in the composites [11]. The hollow structure of plant fibers further facilitates the absorption of water molecules of bio-composites. Several studies demonstrate that bio-composites can achieve a high weight increase of over 10% within few months immersed into water [12,13].

A degradation of bio-composites in mechanical properties caused by water absorption is reported as well. Modulus was pointed out to decrease significantly [14]. Fiber–matrix interface weakening and plasticization phenomena were found to be the main damage mechanisms induced by water ageing [15].

Apparently, the abovementioned moisture sensitivity of bio-composites in mechanical properties should be overcome to further extend its applications in structural components. Fiber treatments are promising to address the moisture absorption issue if fiber-matrix bonding is improved [6]. Among the proposed fiber treatments on plant fibers up to now, influence of the mechanical properties of composites was mainly reported [16]. Only few focused on the influence of treatments on the moisture absorption properties of composites and the effects of treatment on preservation of mechanical properties after moisture absorption. Among them, Gassan et al. [17] produced jute fiber epoxy composites and obtained a reduced water uptake of 10–20% by applying silane treatment on the jute fibers. A reduction of moisture effects on mechanical properties was pointed out. Kushwaha and Kumar [18] investigated the effect of silane treatments on the water absorption properties of bamboo matting reinforced epoxy composites. A decrease in water uptake was obtained, from 41% to 24%. Similarly, Zahari et al. [19] revealed that silane treated ijuk fiber composites had not much difference in water absorption compared to the untreated counterparts. Zhu et al. [20] performed five commonly used pretreatments on flax fibers; water resistance of composites with treated fibers were not improved as expected, except for a slight improvement by alkaline treatment.

Hence, fiber pre-treatment is a promising way to reduce moisture absorption of natural composites. However, further investigations on fiber treatments are still required to improve moisture resistance of natural fiber composites and to better understand the effects of treatments on the water absorption of composites. Furfuryl alcohol (FA) therefore came into our view, being a promising reactant material for treatments on natural fibers. FA is considered as a green chemical because it is mass produced from green renewable agricultural wastes [21,22]. Apart from the environmental merit, it has several advantages for treatments, such as low viscosity and low molecular weight, which allow in-situ polymerization after penetrating into fiber cells. Applications of furfuryl alcohol can already be found in wood preservation to enhance the moisture resistance and dimensional stability of cellulosic woods [23–25]. For natural fiber reinforced composites, few studies are available on the manufacture full bio-composites with polymerized furfuryl alcohol (PFA) as the matrix, while problems of compatibility and solvent removal remain to be addressed [21,26,27]. As for fiber pretreatment with FA, Saw et al. [28] conducted a study in which luffa cylindrical fiber was grafted with FA followed by oxidation. Thermal and mechanical properties of composites was reported to be improved. However, to our knowledge, very few studies have applied treatment on flax fibers with FA. An alternative green fiber pretreatment on flax fibers with FA is then proposed, aiming to improve the water resistance of flax fiber composites used in structural applications.

This paper aims to investigate a green FA pretreatment on flax fibers and its influence on the mechanical and retention properties of flax fiber composites upon moisture absorption. Two different parameters were chosen to investigate the influence of heating cycles on the effects of treatment. Changes in fiber bundle strength caused by treatments were highlighted. Unidirectional composites reinforced with treated and untreated flax fibers were manufactured by vacuum assistant resin transform molding. Moisture resistance of composites was evaluated by water absorptions tests. Specimens were immersed into distilled water bath for different durations to obtain different moisture contents. Mechanical properties of composites with three different moisture contents were then analyzed to compare the influence of treatment on the retention properties of composites.

2. Materials and Methods

2.1. Bio-Based Epoxy Matrix and Flax Fibers

The selected matrix was a two-component green epoxy system (InfuGreen 810/SD 8822, mixing ratio in weight 100:31, from Sicomin, Châteauneuf les Martigues, France). The resin (InfuGreen 810, Sicomin, Châteauneuf les Martigues, France) was derived from colza and has a carbon green content of 38%. The used flax fibers were in the form of non-crimp unidirectional fabric (Bcomp Ltd., Fribourg, Switzerland). Fibers were twisted together into yarns and snitched with textured polyester (1/cm). The fabric had a weight of 300 ± 5% g/m^2. Furfuryl alcohol is a laboratory chemical obtained from Alfa Aesar (supplier: Thermo Fisher GmbH, Karlsruhe, Germany). Applied catalyst for polymerization of furfuryl alcohol was p-Toluenesulfonic acid (PTSA), a product from Sigma-Aldrich Chemie GmbH, Steinheim, Germany.

2.2. Furfuryl Alcohol Treatment on Flax Fibers

Furfuryl alcohol solvent was firstly prepared. PTSA was dissolved in distilled water and then mixed with furfuryl alcohol (0.3 g PTSA and 10 g water for 100 g FA). Flax fabrics in desired sizes (280 mm × 270 mm) were impregnated with FA solvent. Content of FA absorbed by flax fabrics was carefully controlled at 40% ± 1% by using paper towels to absorb excessive FA solvent. Afterwards the impregnated fabrics were wrapped with aluminum foils and placed into oven. Two different heating cycles were used to investigate the effects of temperature on the treatments (Table 1). Treatment 2 was selected as the treatment parameters to investigate the influence of FA treatment on the mechanical properties at different moisture contents. Specimens that underwent Treatment 1 were not evaluated at every moisture content. Only tensile properties before and after 243 h immersed in water bath (also only part of moisture absorption tests) were investigated for Treatment 1 to analyze whether a different heating cycle would impose a significant influence on the effects of FA treatment. It should also be noted that the selected high temperatures in this work were determined by Malte Thomsen, who worked under the author's supervision to find satisfactory treatment parameters in preliminary investigations.

Table 1. Heating cycles of furfuryl alcohol (FA) treatments.

Treatments	Heating Cycles
T1	120 °C for 6 h → 150 °C for 4.5 h
T2	120 °C for 2 h → 150 °C for 7 h

2.3. Composite Manufacturing

Composites were manufactured through vacuum assisted resin transfer molding. Vacuum was applied during the resin infusion process to assure a low porosity level. Pressurized air (2 bar) was applied after resin infusion during the whole curing cycle (30 °C for 8 h followed by 60 °C for 16 h). The resulting composites have a fiber volume fraction of 40%, calculated by surface density of flax fabrics and density of flax fibers.

2.4. Water Absorption

Water absorption was conducted according to ISO-62 [29]. Samples were cut (dry cutting with air as the coolant) into sizes for the tensile tests detailed in the following section. Edges of specimen were carefully polished using a two-step process (grit sizes of sand paper: 320, 1000) to eliminate the influence of edge quality on water uptake. Polished specimens were then conditioned in a desiccator for 12 h at controlled temperature (20 °C) and relative humidity (35% RH). They were then immersed into distilled water bath (room temperature) for 40 h, 115 h and 243 h to obtain different moisture contents. Water absorption behavior was studied by investigating the weight change of the specimens. It was calculated using the following equation:

$$C_t = \frac{m_2 - m_1}{m_1} \times 100 \quad (1)$$

where C_t is weight change of a specimen after an arbitrary time interval t being immersed; m_2 is the mass of tested specimen after immersion; and m_1 is the initial mass of tested specimen after conditioning in desiccator.

2.5. Mechanical Tests on Composites

Tensile properties in fibers direction (0°) and perpendicular to fiber direction (90°) were tested according to ASTM 3039 [30]. Load capacity levels of test machines and basic test set-ups are listed in Table 2. Five replicate specimens were tested for each treatment condition, except that four specimens were tested in the case of 0° specimen immersed for 243 h. For 0° specimens, Young's modulus was defined via the secant modulus between two strain points: 0.05% and 0.25%. For 90° specimens, the same method was used to calculate the transverse modulus, expect that the chosen two strain points are 0.2% and 0.3% to avoid the pronounced slack found at the initial loading stage. Strain data of 90° specimen were obtained from the displacement of traverse. The used sample clamp and testing machine were considered as ideally stiff. However, the transverse modulus calculated in this work is lower than the value predicted by role of mixture. Therefore, the data might not reflect the true intrinsic properties of the composites. Typical problem of grip compliance could be responsible for the deviation. Hence the apparent transverse modulus is then used for phrasing in this paper, instead of transverse modulus. Since the same clamp was used for all tested samples, comparisons of apparent transverse modulus are still considered to be valid.

Table 2. Tensile tests set-ups.

Specimen	Testing Machine	Specimen Dimensions (mm)	Strain Measurement	Testing Speed
0°	Zwick 100 kN	2 × 18 × 250	Clip-on extensometer	2 mm/min
90°	Zwick 2.5 kN	2 × 20 × 110	Displacement of traverse	1 mm/min

2.6. Tensile Tests on Flax Fiber Bundles (Yarns)

Individual flax fiber bundles were extracted directly from flax fabrics at random. Paper tapes were wrapped at both ends of yarn for load introduction. The distance between tape ends was 100 mm. All yarns were loaded at a speed of 5 mm/min until breakage. It should be noted that the cross section area of yarns was not measured because of the large scatter along with one individual yarn and the irregular shape of the cross section. Only the breaking forces were evaluated. At least 10 yarns were extracted from for each type of fabrics at random and then tested.

2.7. Scanning Electron Microscope (SEM)

In this paper, SEM observation was conducted on Phenom XL (Thermo Fisher Scientific, Phenom-World B.V., Eindhoven, The Netherlands). Secondary Electron Detection (SED) was chosen under an acceleration voltage of 10 kV for high quality pictures. All samples were sputtered a thin layer of gold in thickness of about 6.5 nm.

2.8. Fourier Transform Infrared Spectroscopy (FTIR)

FTIR analysis was applied to investigate the polymerization degree of FA. Tests were performed on Bruker TENSOR II (Bruker Corporation, Billerica, MA, USA) with an ATR-unit (Attenuated Total Reflection). The chosen range was from 500 cm^{-1} to 4000 cm^{-1}. The resolution was determined to be 4 cm^{-1} and the spectra were averaged over 16 scans. Spectra from five different locations on each treated or untreated flax fabric were measured, and the average spectra were used.

3. Results

3.1. Characterization of FA Treatment

Impregnated FA in the flax fabrics polymerized into PFA after the heating treatments. The color of flax fabric turned from off white to yellow-red (Figure 1) after FA treatment. Such change of color is ascribed to the formation of chromophore during polymerization process of FA [22,27]. Weight gain of flax fabrics after FA Treatment 2 is about 10%, while flax fabrics that underwent Treatment 1 have a slightly higher weight gain of about 11%.

(a) (b)

Figure 1. (**a**) Untreated flax fiber fabric; and (**b**) furfuryl alcohol (FA) treated flax fiber fabric.

A closer observation on the influence of FA treatment on fiber surface morphology was achieved by SEM. As shown in Figure 2a, most areas of untreated flax fibers surfaces exhibit a smooth appearance, with only some degraded regions probably caused by over retting during flax fiber production. In contrast, parts of treated flax fiber surfaces were covered by thin layers of winkle materials, which can be recognized as PFA (Figure 2b).

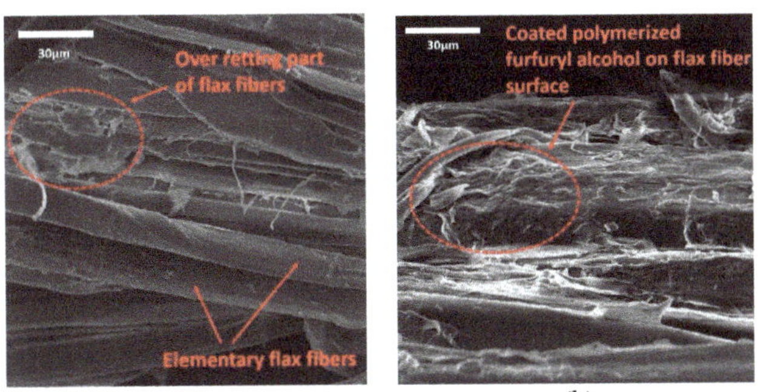

(a) (b)

Figure 2. (**a**) Morphology of untreated flax fiber surfaces; and (**b**) morphology of FA treated flax fiber surfaces.

FTIR spectra of treated flax fabric also give indication of presence of PFA on the flax fiber surfaces (Figure 3). The broad band from 3000 cm^{-1} to 3600 cm^{-1} shows a fading after the FA treatment, owing to the reduced hydroxyl groups on the flax fiber surfaces which are covered by PFA coating. Peaks at 1704 cm^{-1} and 1504 cm^{-1}, which are the feature peaks of the furan rings [21,26], indicate the presence of PFA. No significant difference was observed between spectra of two treated flax fabrics.

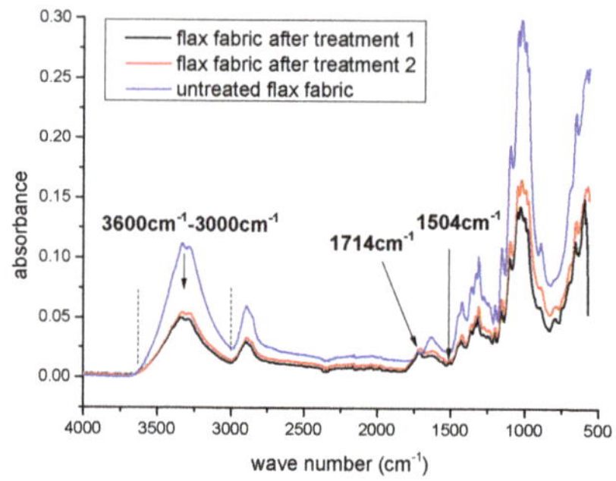

Figure 3. Infrared spectrum of FA treated and untreated flax fabrics.

3.2. Tensile Tests on Flax Fiber Bundles

Figure 4 shows representative force-elongation curves from tensile tests of flax fiber bundles. All types of yarns exhibit brittle tensile behavior. FA treated yarns tended to have a considerably stiffer response upon tensile loads. However, a weakening effect was observed for treated yarns. As summarized in Table 3, the strength of yarns was significantly reduced by FA treatments. Treated yarns broke at lower load levels and resist much less deformation before fracture comparing to untreated yarns. No significant difference was observed between two treatments.

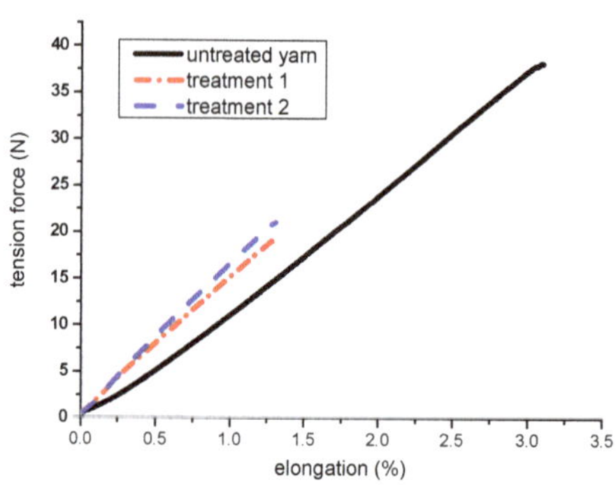

Figure 4. Representative tensile force-elongation curves of treated and untreated yarns.

Table 3. Tensile test results of FA treated and untreated yarns (average and standard deviation).

Treatments	Break Force (N)	Strain at Breakage (%)
untreated	37.16 ± 2.94	2.94 ± 0.43
T1	19.76 ± 6.75	1.108 ± 0.356
T2	20.17 ± 5.49	1.171 ± 0.36

3.3. Moisture Absorption Tests

Results from moisture absorption tests are summarized in Table 4. It can be seen that composites reinforced with FA treated fibers show superior properties of moisture absorption reduction comparing to their untreated counterparts. Moreover, an improvement in moisture resistance was observed over the whole investigated time span (40 h, 115 h and 243 h), as displayed in Table 4. Water absorption measured by weight increase was higher for 90° specimens for both treated and untreated composites.

Table 4. Weight increase (%) of FA treated and untreated composites after water absorption (average and standard deviation).

Time (h)	0° Samples			90° Samples		
	Untreated	T1	T2	Untreated	T1	T2
40	0.630 ± 0.042	0.405 ± 0.170	0.415 ± 0.144	1.294 ± 0.209	0.569 ± 0.093	0.593 ± 0.082
115	1.358 ± 0.144	-	0.829 ± 0.166	3.199 ± 0.321	-	1.311 ± 0.135
243	2.414 ± 0.170	1.389 ± 0.119	1.380 ± 0.093	5.204 ± 0.337	-	2.755 ± 0.442

3.4. *Tensile Tests on Flax Fiber Composites*

Tensile tests results are summarized in Tables 5 and 6. Different responses after FA treatments were observed in 0° and 90° specimen. The tensile strength in fiber direction (longitudinal tensile strength) was observed to decrease after FA treatment, while the longitudinal Young's modulus increased. On the contrary, both the tensile strength and modulus perpendicular to fiber direction (transverse) demonstrate higher values for FA treated ones.

Table 5. Tensile properties in fiber direction of FA treated and untreated composites after water absorption (average and standard deviation).

Time (h)	Untreated		T2		T1	
	Strength (MPa)	Young Modulus (GPa)	Strength (MPa)	Young Modulus (GPa)	Strength (MPa)	Young Modulus (GPa)
0	290.00 ± 3.08	20.17 ± 0.52	196.40 ± 4.62	23.62 ± 0.38	200.6 ± 17.37	23.8 ± 1.33
40	271.20 ± 12.60	17.32 ± 0.46	194.2 ± 4.02	21.34 ± 0.15	-	-
115	255.00 ± 12.59	15.50 ± 0.25	186.00 ± 6.89	20.42 ± 0.25	-	-
243	265.3 ± 15.13	13.46 ± 0.47	180.00 ± 5.58	20.77 ± 0.06	180.5 ± 2.38	20.4 ± 0.25

Table 6. Tensile properties perpendicular to fiber direction of FA treated and untreated composites after water absorption (average and standard deviation).

Time (h)	Untreated		T2		T1	
	Strength (MPa)	Apparent Modulus (GPa)	Strength (MPa)	Apparent Modulus (GPa)	Strength (MPa)	Apparent Modulus (GPa)
0	25.87 ± 0.92	0.494 ± 0.029	28.36 ± 2.04	0.547 ± 0.026	27.74 ± 1.71	0.545 ± 0.027
40	21.24 ± 0.96	0.425 ± 0.017	24.67 ± 0.97	0.529 ± 0.022	-	-
115	18.20 ± 1.03	0.424 ± 0.020	22.54 ± 0.58	0.526 ± 0.016	-	-
243	18.34 ± 0.81	0.383 ± 0.018	20.41 ± 0.75	0.443 ± 0.011	20.15 ± 0.99	0.462 ± 0.009

Upon water absorption, all tested composites show decreasing trends in both tensile strengths and modulus (longitudinal and transverse). Comparing to untreated composites, tensile properties of treated composites are better retained after moisture absorption.

4. Discussion

4.1. *Influence of FA Treatment on the Mechanical Properties of Composites*

Obviously, the FA treatment imposes a significant effect on the tensile properties of composites. As can be seen in Figure 5, tensile strength in fiber direction is significantly degraded (by about 30%) after FA treatments. Such degradation was also observed in the fiber bundle tensile tests on treated yarns. Break forces of FA treated yarns was found to be reduced by over 45% (Table 3). The consistence between two tests can thus explain the sink of tensile properties in fiber direction. In fact, it has been reported that cellulosic natural fibers can be degraded under acid conditions at high temperatures [31]. Marshall [32] also mentioned a significant strength reduction of flax/cotton nonwoven mat after heating (60 °C and 120 °C) with citric acid. When applying tensile loading in fiber direction, the fibers carry the main load and any weakening of their mechanical properties by degradation would correspondingly result in a decreased maximum load the composites can carry. The treatments in this paper applied a high temperature of 150 °C under strong acid condition, thus can be an important attributing factor to the decrease of tensile properties in fiber direction.

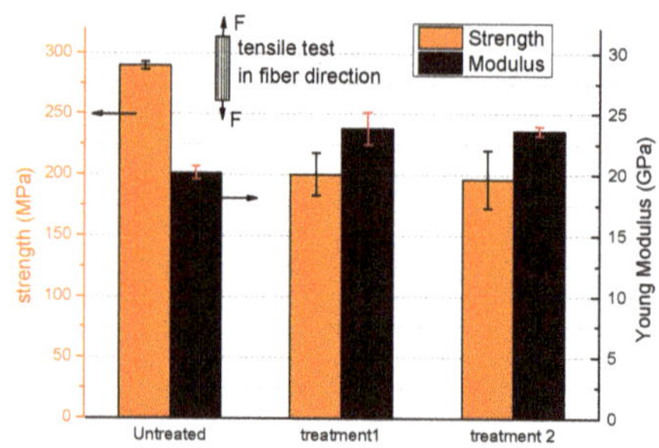

Figure 5. Influence of FA treatments on the tensile properties of composites in fiber direction (error bars indicate standard deviation).

Further investigation on the tensile tests of 90° composites allowed evaluating the effects of treatment on fiber-matrix interfacial bonding properties. As displayed in Table 6, the slight increase in transverse tensile strength and modulus of treated composites could be an indication of improved fiber-matrix bonding properties. In fiber reinforced laminates, good fiber-matrix bonding properties impede the propagation of critical flaw and cracks, resulting in high transverse tensile strengths [33].

The improved fiber-matrix bonding could result from two-fold mechanisms. Firstly, the treated yarns are more compatible with the hydrophobic matrix, owing to the fact that coated PFA on the fiber surfaces increases the hydrophobicity. Secondly, as already shown in Figure 2b, treated fiber surfaces are rougher and offer a larger surface area to interact with matrix. SEM observations on the fracture surface of 90° composites also prove a good fiber-matrix bonding for treated composites. The PFA at interfaces both on and inside a fiber bundle were found to join flax fibers and the matrix strongly, showing no clearance between each other (Figure 6b). Although no conclusive evidence could be given whether there are chemical bonds generated between PFA and matrix, the strong bonding between flax fibers and PFA could be resulted from the formation of covalent bonds through the reactions between PFA chains and hydroxyl groups of lignin on flax fiber surfaces [25,28]. In comparison, debonding between flax fiber bundles and matrix was observed at the interfaces for untreated composites, indicating an inferior bonding between flax fibers and matrix (Figure 6a).

Degradation in strength is a disadvantage of the applied treatment, whereas the Young's modulus of composites in fiber direction is improved by about 18%. This can be partly attributed to a better fiber-matrix bonding as discussed above. A strong fiber-matrix bonding facilitates load transfer from matrix to fibers without causing much deformation in the interfacial regions. On the other hand, coated PFA might have a stiffening effect on loaded fibers (Figure 4). It is assumed that FA penetrates inside the yarns or even possibly in the fiber cells and polymerize into stiff PFA, replacing weak parts of fibers or connections between individual fibers. This would also explain the experimental results for the fiber yarns, but further investigations are necessary to verify this hypothesis.

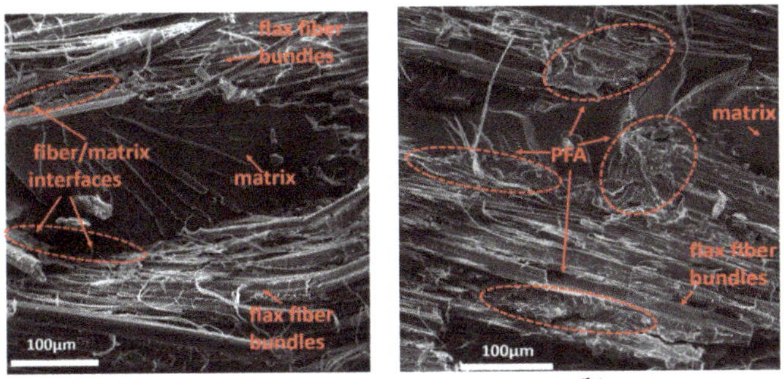

(a) (b)

Figure 6. (**a**) Morphology of fracture surface of untreated flax fiber composites; and (**b**) morphology of fracture surface of FA treated flax fiber composites.

4.2. Influence of FA Treatment on the Moisture Absorption of Composites

Applying FA treatment on flax fibers leads to the expected improvement in moisture absorption retardation. For 0° specimens that underwent FA Treatment 2, the amount of water uptake after 40 h, 115 h and 243 h was reduced by 34%, 39% and 42.4%, respectively, compared with untreated composites (Figure 7). Absorbed water amount of composites after over 10 days reached only 1.389% of original weight, while the untreated composites absorb nearly twice as much water. The improvement of moisture resistance can be ascribed to the better fiber-matrix bonding of the composites which impede the propagation of water molecules along with the fiber-matrix interfaces. Furthermore, flax fibers were more hydrophobic after FA treatment owing to the introduction of PFA on the flax fibers or inside flax fiber cells. In fact, PFA is stated to have an excellent water resistance property. Water uptake of PFA at saturation was reported to have a value of 1.12% [21], which is lower than commonly used polymer matrixes, including the selected bio-epoxy in this work.

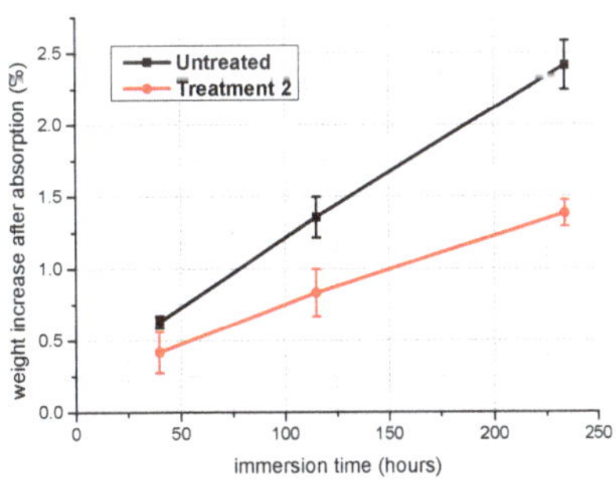

Figure 7. Influence of FA treatments on the water uptake of 0° specimen (error bars indicate standard deviation).

4.3. Influence of FA Treatment on the Retention Properties of Composites after Water Absorption

Retention of mechanical properties in wet conditions is vitally important for composites used in structural applications. Severe degradation of wet composites in mechanical properties is the major reason moisture resistance of composites is required to be improved.

Figure 8 displays the retention of tensile properties of FA treated specimen (T2) and untreated specimen. The tensile properties of treated composites are better maintained compared to the untreated ones at the same immersion duration. Longitudinal tensile strength of treated composites only decreased by 5.3% after >5 days immersed in water, while their untreated counterparts showed a decrease of 15.5% over the same time span. When it comes to transverse strength, both types of composites show a pronounced decrease along with the immersion time. Untreated composites again demonstrated a worse retention compared to FA treated ones, especially over the first 115 h in water, indicating a faster degradation in fiber-matrix bonding. This is further testified by SEM observations on 90° specimen fracture surfaces. As depicted in Figure 9, there is an evident difference in the fractographic features for untreated composites before and after water absorption, while no obvious change was observed for FA treated composites. Figure 9a depicts the loose fiber bundles/matrix bonding of untreated composites, with flax fibers still bunching up together in the original from of yarns. After absorbing moisture over five days, inter-fiber splitting inside fiber bundles was noticed (Figure 9b), indicating that inter-fiber bonding was weakened significantly by the absorbed water molecules. In contrast, FA treated composites show similar fractographic features of tight flax fiber-matrix bonding and broken fibers bunches before and after 115 h in water, revealing a better resistance of fiber-matrix bonding against water uptake.

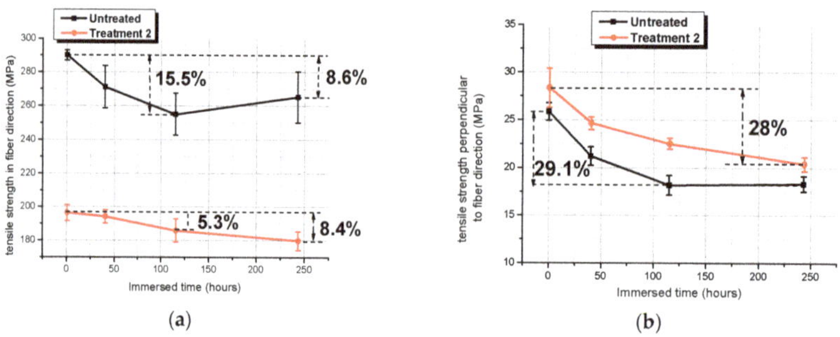

Figure 8. Comparison of retention rates in tensile strength after water uptake: (**a**) comparison of longitudinal tensile strength between treated and untreated composites; and (**b**) comparison of transverse tensile strength between treated and untreated composites. (Error bars indicate standard deviation).

Figure 10a shows the retention of Young's modulus in fiber direction, which is an indication of load transfer efficiency at fiber-matrix interfaces and intrinsic fiber tensile modulus. It is clearly revealed that FA treatment on composites demonstrated not only a significantly increased stiffness over the untreated composites (Table 5), but also reserve larger extent of Young's modulus. Stronger fiber-matrix bonding again explains the better resistance on degradation at fiber-matrix interfaces. Meanwhile, the higher hydrophobicity of flax fibers also impeded the plasticizing effect on flax fibers caused by water absorption, which could be responsible for the widening superiority of treatment at longer submersion in water (after 243 h). Similar effects of FA treatment were also observed for transverse modulus (Figure 10b and Table 6), except that the advantage at longer exposure to water was weakened. In fact, the transverse modulus was dominated by the matrix modulus and also fiber-matrix interface stiffness. During the first 115 h in water, degradation in fiber-matrix stiffness of treated composites was less severe for the untreated ones; the treated ones thus maintain a high fiber-matrix stiffness. The matrix stiffness degradation by water absorption could be to some extent compensated. However, it is likely that the compensation effect was weakened upon longer exposure, owing to the fact that both fiber-matrix stiffness and matrix stiffness are further reduced.

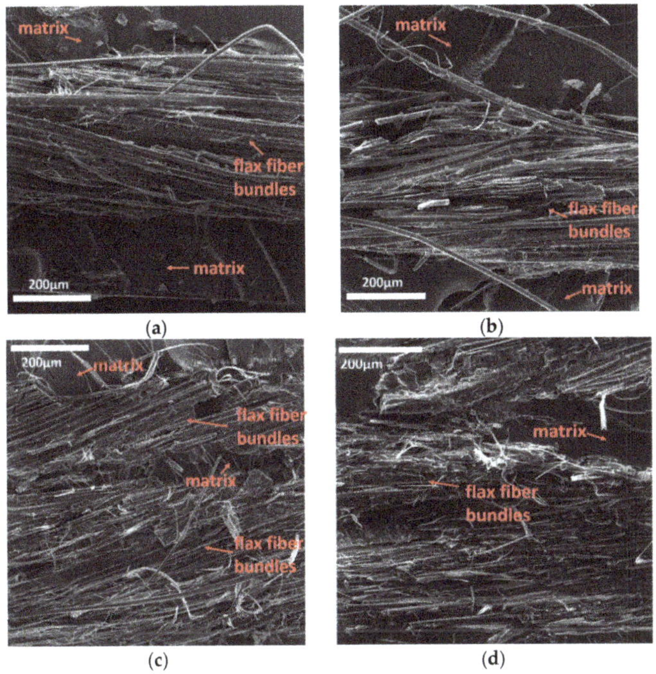

Figure 9. Morphology of fracture surfaces of 90° specimens: (**a**) control group (with untreated flax fibers), 0 h in water; (**b**) control group (with untreated flax fibers), 115 h in water; (**c**) with flax fibers under Treatment 2 (T2), 0 h in water; and (**d**) with flax fibers under Treatment 2 (T2), 115 h in water.

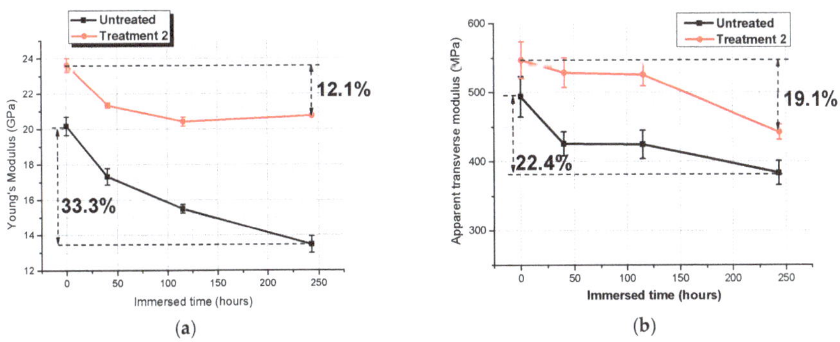

Figure 10. Change of modulus in fiber direction after water absorption: (**a**) retention of Young's modulus in fiber direction; and (**b**) retention of transverse modulus. (Error bars indicate standard deviation).

Interestingly, longitudinal tensile strength of untreated composites after 243 h in water was higher than after 115 h in water (Figure 8), showing a "strengthening" effect upon further water absorption, while the transverse tensile strength remained circa 30% lower. In fact, a high longitudinal tensile strength of unidirectional composites does not necessarily indicate high fiber-matrix bonding [34]. Local stress concentration at broken fiber ends and debonding areas also contribute greatly to the ultimate tensile strength. Under the same fiber-matrix bonding properties, local concentrations in untreated composites (243 h in water) was believed to play a less important role because of further plasticizing effects, proven by a higher failure strain of 2.38% compared to 2.08% after 115 h in water. The higher failure strain also means more movement of fibers to a better alignment in the load direction, partly overcoming the fiber misalignment and thus compensate the weakening fiber-matrix bonding.

4.4. Comparison of FA Treatments with Different Heating Cycles

The polymerization process of FA into PFA is assumed to involve a series of complex reactions [22,27]. It is sensitive to several parameters such as type of catalysts, catalyst contents, temperature histories by polymerization, FA solvent contents of treated fibers, etc. These parameters could affect the influence of FA treatment on flax fiber composites. This paper gives preliminary attention to the influence of two heating cycles on the effects of FA treatment. As shown in Figure 3, the FTIR spectrum of both treated flax fibers show no great difference from each other, which indicates basically similar polymerization degrees. The tensile tests on treated fiber bundles also indicates no clear trend of superiority between the two treatments (Table 3). With regard to the effects on the mechanical properties of composites (Tables 5 and 6) and moisture absorption resistance (Table 4), the discrepancy is also perceived to be not significant. It appears that FA treatment is not sensitive to heating cycles within the selected time and temperature span. However, it should be mentioned that both heating cycles were determined to assure a satisfying polymerization degree of FA. Therefore, it is assumed that FA treatment might not be sensitive to temperature histories, as long as the polymerization degree of FA reaches a certain range when other parameters are held constant. However, further research on the influence of the treatment parameters is still required.

5. Conclusions

This study investigated a fiber pretreatment on flax fiber composites using furfuryl alcohol. Moisture uptake of resulting composites and their retention in tensile properties were analyzed. The results show an expected improvement in moisture resistance and a better retention of tensile strength and stiffness for treated composites. FA treatment yielded an up to 18% improved modulus, and the moisture uptake was reduced significantly. The fiber-matrix interfacial bonding was also found to be improved. During the FA treatment in parallel, the acid catalyst reduced the strength of the flax fibers. It was shown that the FA pretreatment is an efficient way to reduce the moisture absorption. Further work has to focus on the catalyst (acid) and treatment process to prevent degradation of the fibers.

Author Contributions: Y.J. and B.F. proposed the idea of FA treatment and conceived the experiments. Y.J. carried out the experiments, data analyzing and wrote the paper. B.F. reviewed and revised the paper.

Funding: This research received no external funding. The publication was supported by the Deutsche Forschungsgemeinschaft (DFG, German Research Foundation)—project number 392323616 and the Hamburg University of Technology (TUHH) in the funding program "Open Access Publishing".

Acknowledgments: Y.J. would like to express his gratitude to China Scholarship Council (CSC) for providing a PhD scholarship. The authors thank Hakan Özdemir for his contribution in part of the experimental work and Malte Thomsen for his preliminary investigation on the FA treatment process.

Conflicts of Interest: The authors declare no conflict of interest.

References

1. Bodros, E.; Pillin, I.; Montrelay, N.; Baley, C. Could biopolymers reinforced by randomly scattered flax fibre be used in structural applications? *Compos. Sci. Technol.* **2007**, *67*, 462–470. [CrossRef]
2. Faruk, O.; Bledzki, A.K.; Fink, H.-P.; Sain, M. Progress Report on Natural Fiber Reinforced Composites. *Macromol. Mater. Eng.* **2014**, *299*, 9–26. [CrossRef]
3. Shah, D.U.; Schubel, P.J.; Clifford, M.J. Can flax replace E-glass in structural composites? A small wind turbine blade case study. *Compos. Part B Eng.* **2013**, *52*, 172–181. [CrossRef]

4. Goutianos, S.; Peijs, T.; Nystrom, B.; Skrifvars, M. Development of Flax Fibre based Textile Reinforcements for Composite Applications. *Appl. Compos. Mater.* **2006**, *13*, 199–215. [CrossRef]
5. Ku, H.; Wang, H.; Pattarachaiyakoop, N.; Trada, M. A review on the tensile properties of natural fiber reinforced polymer composites. *Compos. Part B Eng.* **2011**, *42*, 856–873. [CrossRef]
6. Pickering, K.L.; Efendy, M.A.; Le, T.M. A review of recent developments in natural fibre composites and their mechanical performance. *Compos. Part A Appl. Sci. Manuf.* **2016**, *83*, 98–112. [CrossRef]
7. Jauhari, N.; Mishra, R.; Thakur, H. Natural Fibre Reinforced Composite Laminates–A Review. *Mater. Today* **2015**, *2*, 2868–2877. [CrossRef]
8. Akampumuza, O.; Wambua, P.M.; Ahmed, A.; Li, W.; Qin, X.-H. Review of the applications of biocomposites in the automotive industry. *Polym. Compos.* **2017**, *38*, 2553–2569. [CrossRef]
9. Zini, E.; Scandola, M. Green composites: an overview. *Polym. Compos.* **2011**, *32*, 1905–1915. [CrossRef]
10. Shekar, H.S.; Ramachandra, M. Green Composites: A Review. *Mater. Today* **2018**, *5*, 2518–2526. [CrossRef]
11. Mokhothu, T.H.; John, M.J. Review on hygroscopic aging of cellulose fibres and their biocomposites. *Carbohydr. Polym.* **2015**, *131*, 337–354. [CrossRef] [PubMed]
12. Hristozov, D.; Wroblewski, L.; Sadeghian, P. Long-term tensile properties of natural fibre-reinforced polymer composites: Comparison of flax and glass fibres. *Compos. Part B Eng.* **2016**, *95*, 82–95. [CrossRef]
13. Cheour, K.; Assarar, M.; Scida, D.; Ayad, R.; Gong, X.-L. Effect of water ageing on the mechanical and damping properties of flax-fibre reinforced composite materials. *Compos. Struct.* **2016**, *152*, 259–266. [CrossRef]
14. Assarar, M.; Scida, D.; El Mahi, A.; Poilâne, C.; Ayad, R. Influence of water ageing on mechanical properties and damage events of two reinforced composite materials: Flax–fibres and glass–fibres. *Mater. Des.* **2011**, *32*, 788–795. [CrossRef]
15. Chilali, A.; Zouari, W.; Assarar, M.; Kebir, H.; Ayad, R. Effect of water ageing on the load-unload cyclic behaviour of flax fibre-reinforced thermoplastic and thermosetting composites. *Compos. Struct.* **2018**, *183*, 309–319. [CrossRef]
16. Kabir, M.M.; Wang, H.; Lau, K.T.; Cardona, F. Chemical treatments on plant-based natural fibre reinforced polymer composites: An overview. *Compos. Part B Eng.* **2012**, *43*, 2883–2892. [CrossRef]
17. Gassan, J.; Bledzki, A.K. Effect of moisture content on the properties of silanized jute-epoxy composites. *Polym. Compos.* **1997**, *18*, 179–184. [CrossRef]
18. Kushwaha, P.K.; Kumar, R. Studies on water absorption of bamboo-epoxy composites: Effect of silane treatment of mercerized bamboo. *J. Appl. Polym. Sci.* **2010**, *115*, 1846–1852. [CrossRef]
19. Zahari, W.; Badri, R.; Ardyananta, H.; Kurniawan, D.; Nor, F.M. Mechanical Properties and Water Absorption Behavior of Polypropylene/Ijuk Fiber Composite by Using Silane Treatment. *Procedia Manuf.* **2015**, *2*, 573–578. [CrossRef]
20. Zhu, J.; Zhu, H.; Abhyankar, H.; Njuguna, J. Effect of Fibre Treatments on Water Absorption and Tensile Properties of Flax/Tannin Composites. In Proceedings of the 19th International Conference on Composite Materials, Montreal, QC, Canada, 28 July–2 August 2013.
21. Deka, H.; Misra, M.; Mohanty, A. Renewable resource based "all green composites" from kenaf biofiber and poly(furfuryl alcohol) bioresin. *Ind. Crops Prod.* **2013**, *41*, 94–101. [CrossRef]
22. Choura, M.; Belgacem, N.M.; Gandini, A. Acid-catalyzed polycondensation of furfuryl alcohol: Mechanisms of chromophore formation and cross-linking. *Macromolecules* **1996**, *29*, 3839–3850. [CrossRef]
23. Li, W.; Wang, H.; Ren, D.; Yu, Y.; Yu, Y. Wood modification with furfuryl alcohol catalysed by a new composite acidic catalyst. *Wood Sci. Technol.* **2015**, *49*, 845–856. [CrossRef]
24. Lande, S.; Westin, M.; Schneider, M. Properties of furfurylated wood. *Scand. J. For. Res.* **2004**, *19*, 22–30. [CrossRef]
25. Nordstierna, L.; Lande, S.; Westin, M.; Karlsson, O.; Furó, I. Towards novel wood-based materials: Chemical bonds between lignin-like model molecules and poly(furfuryl alcohol) studied by NMR. *Holzforschung* **2008**, *62*, 267. [CrossRef]
26. Motaung, T.E.; Gqokoma, Z.; Linganiso, L.Z.; Hato, M.J. The effect of acid content on the poly(furfuryl) alcohol/cellulose composites. *Polym. Compos.* **2016**, *37*, 2434–2441. [CrossRef]
27. Crossley, R.J.; Schubel, P.J.; Stevenson, A.; Moreira, M. The Development and Processing of a Sustainable Fully bio Derived Polyfurfuryl Alcohol Matrix Flax Fibre Prepreg. In Proceedings of the 15th European Conference on Composite Materials, Venice, Italy, 24–28 June 2012.

28. Saw, S.K.; Purwar, R.; Nandy, S.; Ghose, J.; Sarkhel, G. Fabrication, Characterization, and Evaluation of Luffa cylindrica Fiber Reinforced Epoxy Composites. *BioResources* **2013**, *8*. [CrossRef]
29. *BS EN ISO. 62. Plastics–Determination of Water Absorption 1999*; American National Standards Institute: New York, NY, USA, 2007.
30. ASTM. *D3039/D3039M. Standard Test Method for Tensile Properties of Polymer Matrix Composite Materials*; American Society for Testing and Materials: West Conshohocken, PA, USA, 2000.
31. Wyman, C.E. Ethanol from lignocellulosic biomass: technology, economics, and opportunities. *Bioresour. Technol.* **1994**, *50*, 3–15. [CrossRef]
32. Marshall, W.E.; Akin, D.E.; Wartelle, L.H.; Annis, P.A. Citric acid treatment of flax, cotton and blended nonwoven mats for copper ion absorption. *Ind. Crops Prod.* **2007**, *26*, 8–13. [CrossRef]
33. Liebig, W.V.; Leopold, C.; Hobbiebrunken, T.; Fiedler, B. New test approach to determine the transverse tensile strength of CFRP with regard to the size effect. *Compos. Commun.* **2016**, *1*, 54–59. [CrossRef]
34. Subramanian, S.; Lesko, J.J.; Reifsnider, K.L.; Stinchcomb, W.W. Characterization of the fiber-matrix interphase and its influence on mechanical properties of unidirectional composites. *J. Compos. Mater.* **1996**, *30*, 309–332. [CrossRef]

fibers

Article

Mechanical Properties of a Water Hyacinth Nanofiber Cellulose Reinforced Thermoplastic Starch Bionanocomposite: Effect of Ultrasonic Vibration during Processing

Mochamad Asrofi [1], Hairul Abral [1,*], Anwar Kasim [2], Adjar Pratoto [1], Melbi Mahardika [1] and Fadli Hafizulhaq [1]

1 Department of Mechanical Engineering, Andalas University, 25163 Padang, Sumatera Barat, Indonesia; asrofi.net@gmail.com (M.A.); adjar.pratoto@ft.unand.ac.id (A.P.); melbimahardika@gmail.com (M.M.); hafizulhaq.fadli@gmail.com (F.H.)

2 Department of Agriculture Technology, Andalas University, 25163 Padang, Sumatera Barat, Indonesia; anwar_ks@yahoo.com

* Correspondence: abral@ft.unand.ac.id; Tel.: +62-812-672-0355

Received: 12 March 2018; Accepted: 5 June 2018; Published: 8 June 2018

Abstract: Thermoplastic starch (TPS) reinforced by 1 wt % nanofiber cellulose (NFC) reinforcing from water hyacinth was produced. Ultrasonic vibration time (UVT) was applied to bionanocomposites during gelation for 0, 15, 30 and 60 min. Morphology of the NFC was investigated using Transmission Electron Microscopy (TEM). Scanning Electron Microscopy (SEM) and tensile tests were performed to identify the fracture surface and determine the mechanical properties of the bionanocomposites, respectively. The Crystallinity index (CI) of untreated and treated bionanocomposites was measured using X-ray Diffraction (XRD). The average diameter of NFC water hyacinth was 10–20 nm. The maximum tensile strength (TS) and modulus elasticity (ME) of the bionanocomposite was 11.4 MPa and 443 MPa respectively, after 60 min UVT. This result was supported by SEM which indicated good dispersion and compact structure.

Keywords: nanofiber cellulose; water hyacinth; thermoplastic starch; bionanocomposites; ultrasonic vibration time

1. Introduction

Reducing the use of synthetic plastics has become a special focus for researchers due to the environmental problems caused by non-biodegradable plastic waste. Recent innovations have included mixing synthetic with bio-based material (hybrid) and using fully biodegradable materials (bioplastics). The development of bioplastics to replace synthetic plastics has gained prominence in the last 10 years [1,2]. Bioplastics can be made from starch using shear stress and heat treatment at 60–70 °C [3,4] and they can be low cost and environmental friendly [5,6]. However, these plastics have various disadvantages including lower mechanical strength, water resistance, and thermal properties than synthetic plastics [7,8]. These disadvantages can be reduced by addition of natural fiber as a reinforcing agent into the starch matrix. The natural fibers commonly used are microfibers, with a diameter in micron meter area (1–1000 μm), or nanofibers with a diameter of 1–100 nanometers [9,10]. Microfibers sourced from kenaf [4], water hyacinth [5,11], ramie [12,13], empty palm oil bunches (TKKS) [6], and microalgae [14] and chitin [15] have been used in previous studies. However, these microfibers usually form agglomerations and porosities when dispersed in a starch matrix and resulting in low mechanical properties [6].

To reduce the agglomeration, fiber must be made nano-sized. Smaller fibers result in better biocomposites due to high contact surface area, lower density, and good mechanical properties [9,16]. However, even nanofiber may not distribute homogeneously throughout the starch matrix. This may result in the reduction of the mechanical properties of the biocomposite [17]. Furthermore, nanofiber is expensive to produce and processing often results in low nanofiber yield. This is a problem as generally higher loading values of fiber are necessary for significant improvements in properties of the resulting biocomposite material [18].

Ultrasound treatment after gelation has been recommended to overcome this as it uniformly distributes the fiber throughout the matrix as the kinetic energy is released by ultrasound, which breaks down the fiber agglomerations into individual fibers scattered throughout the matrix [6,19]. This has been reported to result in more compact structures, improved mechanical properties, water resistance, and good film formation [5,6,20].

Water hyacinth is one possible source of natural fiber which is low cost, non-toxic, and abundant in Indonesia. It has approximately 40% cellulose content which is comparable with the fiber content from other sources [21]. Cellulose content has an important influence on the properties of biocomposites. The higher cellulose content leads to better mechanical and thermal properties as reported in previous study [22,23].

Yam bean is a common crop in tropical countries like Indonesia and can be harvested every three months. Yam bean starch has a particularly high amylose content in the range of 30–40% [8]. This is higher than in tapioca (15%) [5], corn (28%) [14], and potato starch (25%) [24]. Previous research has shown that amylose-rich starch results in bioplastics with superior mechanical properties [8,25].

Therefore, we chose to investigate the properties of a biocomposite manufactured from a low loading of nanofiber cellulose (NFC) from water hyacinth in a TPS matrix from yam beans (YB) in this study. To attempt to produce a bioplastic free from agglomerations and hence even better properties, we used ultrasonic vibration during gelation. To date, there are no published studies on the effect of UVT during the gelation of TPS from YB starch reinforced by water hyacinth NFC. Morphological, mechanical properties and crystallinity index have been analyzed in this study.

2. Materials and Methods

2.1. Materials

Water hyacinth fiber was obtained from a river in Payakumbuh, Indonesia. The content of cellulose, hemicellulose, and lignin was 43%, 29%, and 7%, respectively. Yam bean (YB) starch (43% amylose and 57% amylopectin), glycerol (density: 1.255–1.260 g/mL; Brataco brand), and distilled water were supplied by the Mechanic Metallurgy Laboratory, Andalas University.

2.2. NFC Preparation

The water hyacinth NFC preparation was carried out as in [21] with certain modifications in the mechanical treatment. In this study, we used ultrasonic crusher treatment to obtain the NFC. The cellulose suspension was sonicated using ultrasonic cell crusher (SJIA-1200W) at 600 W for 1 h after chemical treatment. The temperature of the cellulose suspension was kept below 50 °C.

2.3. Production of TPS Bionanocomposites

A mixture of 2 mL glycerin, 100 mL distilled water, 10 wt % YB starch and 1 wt % NFC was heated on a hot plate in a glass beaker with a magnetic stirrer at 60 °C, 500 rpm for 25–30 min until gelation. During gelation, the bionanocomposite gel was treated in an ultrasonic bath (model number PS-70AL, 40 kHz, and 420 W). The treatment was conducted at 40 kHz, 250 W, and 100% amplitude on four different samples with UVT of 0, 15, 30 and 60 min. The water temperature in the bath was kept under 40 °C. The gel was cast in a petri dish (diameter: 15 cm) and dried in a Memert UN50 oven at 50 °C for 20 h.

2.4. Characterization

2.4.1. Morphology of Untreated and Treated Water Hyacinth Fiber

The morphology of raw (untreated), bleached, and acid hydrolysis fiber was observed by using the SEM Hitachi 3400 N instrument. It was observed at 10 kV and room temperature.

2.4.2. Morphology of NFC Water Hyacinth

The fiber was studied 1 h after sonication using TEM (JEOL JEM-1400). The fiber suspension was dripped onto a copper grid coated by carbon film and dried at room temperature. It was observed using TEM at 100 keV.

2.4.3. Fracture Surface Bionanocomposites

The fracture surface of all bionanocomposite films after the tensile test was observed by SEM VEGA3 TESCAN at room temperature. The operation voltage was HV 10 kV. Before characterization, all films were coated with Palladium-Gold (Pd-Au).

2.4.4. Mechanical Properties

Tensile tests were performed to determine the mechanical properties of the bionanocomposite films using COM-TEN 95T Series 5K (maximum capacity: 5000 pounds) testing machine. The output of the tensile test was tensile strength, modulus elasticity, and elongation at break. All samples were fabricated in rectangular profile according to American Society for Testing and Materials (ASTM) D882 standard. Length and width overall of sample were 100 and 15 mm, respectively. The gage length was 50 mm. Before testing, all samples tested were conditioned in a desiccator for 2 days (Relative Humidity (RH): 50%). During testing, the speed and RH were maintained at 5 mm/min and 80%, respectively. There were five repeats used for each UVT variation.

Tensile test results were obtained by dividing maximum force by the average cross-sectional area of the sample which was measured at five different points using a micrometer and vernier caliper (accuracy: 0.001 mm) for thickness and width.

2.4.5. Crystallinity Index Calculation

A PANalytical X'pert PRO instrument was used to perform the XRD of all samples tested. The peak intensity was recorded at (2θ = 10–80°) (λ = 0.154 nm) using 40 kV and 35 mA. Crystallinity index (*CI*) of the NFC water hyacinth samples was calculated by the Segal's Equation (1) [26]:

$$CI\ (\%) = \frac{I_{002} - I_{am}}{I_{002}} \times 100 \quad (1)$$

where, I_{002} was maximum crystalline peak intensity at (2θ = 22.6°) and I_{am} the amorphous diffraction intensity at (2θ = 18°). Meanwhile, the *CI* of all bionanocomposites was determined by Hulleman's method [27].

3. Results and Discussion

3.1. Morphology of Untreated and Treated Water Hyacinth Fiber

Figure 1 presents the appearance of raw water hyacinth, bleaching fiber, acid hydrolysis fiber, and water hyacinth NFC processed by ultrasonic cell crusher for 1 h. It can be seen that; raw fiber forms several fiber bundles and has a smooth surface due to the presence of cementing material such as waxes and oils (Figure 1a) [28]. After bleaching, the water hyacinth fiber disintegrated due to the hydrolyzation of hemicellulose and partial removal of the lignin [29]. Acid hydrolysis of the fiber then depolymerized it by breaking up the cellulose chains [18,29]. The resulting fibers have diameters of

1–10 micron. After 1 h sonication (Figure 1d), fibers took the form of short fibrils with average diameter and length of 10–20 nm and 90–250 nm, respectively as has been observed in previous reports [30,31].

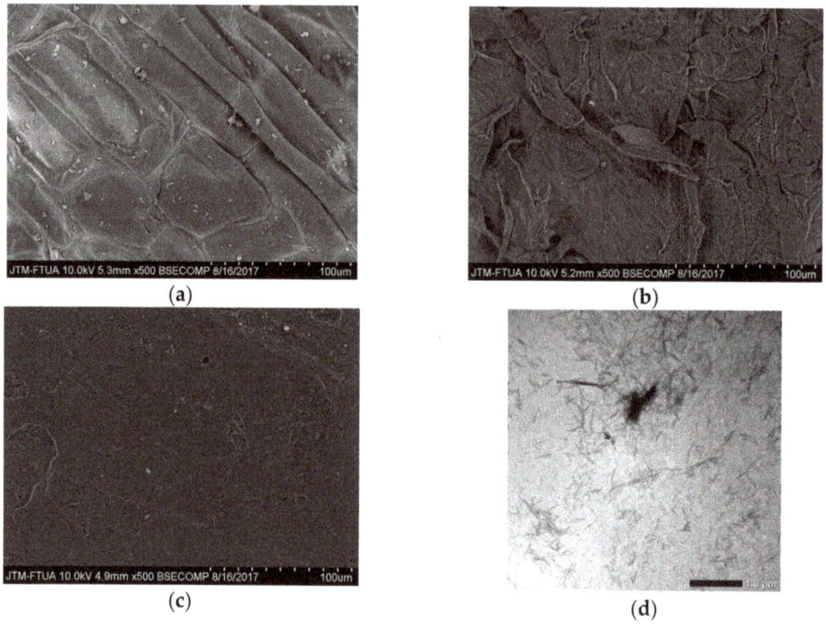

Figure 1. SEM images of morphological: (**a**) raw water hyacinth; (**b**) bleached fiber; (**c**) acid hydrolysis fiber; (**d**) TEM images of nanofiber cellulose (NFC) water hyacinth after sonication for 1 h.

3.2. Fractured Surface of Bionanocomposites

Figure 2 displays SEM imaging of TPS and all fractured bionanocomposite samples after the tensile test. Figure 2a shows the fracture surface of the TPS film. It looks to be a smooth surface and there is no aggregation due to good interaction between starch and glycerol during the fabrication process [32]. The addition of NFC 1 wt % in TPS (untreated) is displayed by Figure 2b,c. As can be seen, there is porosity formation at several spots. The NFC also seems undistributed in starch matrix. This is due to a bad interaction between starch and NFC during fabrication [33]. This would appear to be a cause of the low tensile strength. For 15 min UVT (Figure 2d), the sample NFC agglomerates at one spot due to bad dispersion in the matrix [33,34]. A similar phenomenon was also reported by a previous report [5,6]. After 30 and 60 min UVT, the samples display good nanocellulose dispersion throughout the matrix and compact structure with no porosity or agglomeration formation (Figure 2e,f). The kinetic energy from the ultrasonic bath disperses the agglomerated fiber well resulting in a good distribution throughout the matrix leading to good mechanical properties [6,20,35].

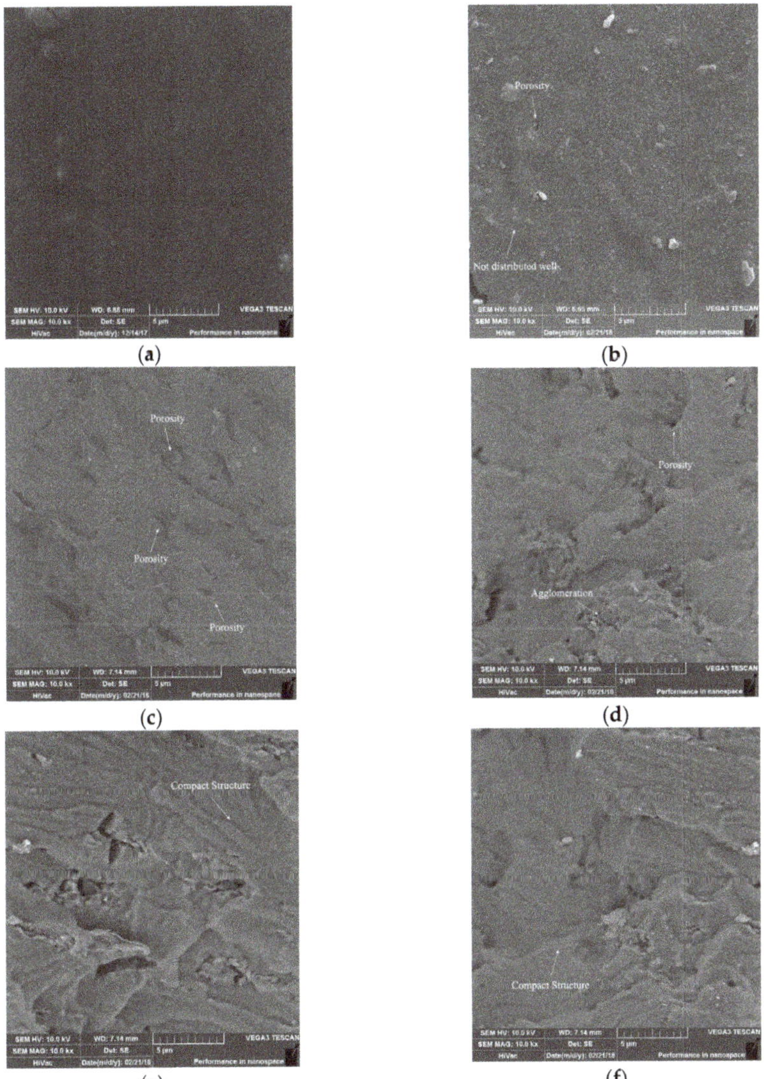

Figure 2. SEM images of fractured surface: (**a**) pure thermoplastic starch (TPS) film; (**b**) TPS reinforced 1 wt %; bionanocomposites with different UVT: (**c**) 0 min (untreated); (**d**) 15 min; (**e**) 30 min; and (**f**) 60 min.

3.3. Mechanical Properties

The effect of UVT on mechanical properties of bionanocomposite is displayed in Figure 3. The TS and ME increased significantly with longer UVT. The 0 min UVT (untreated) bionanocomposite (Figure 3a) had a TS of 4 MPa. This result was lower than in the treated bionanocomposite due to porosities formation and poorly distributed fiber in the starch matrix (see Figure 2b,c). After 15 min UVT, the TS was increased by 45%. Longer UVT for 30 and 60 min increased TS to 82% and 185%, respectively.

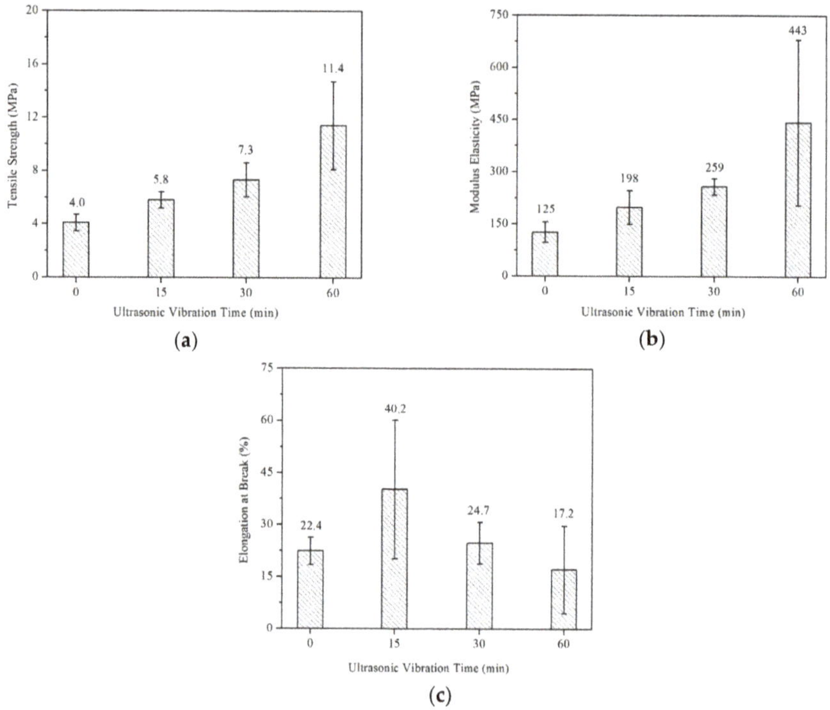

Figure 3. Effect ultrasonic vibration time (UVT) on: (**a**) Tensile Strength (TS); (**b**) Modulus Elasticity (ME); and (**c**) Elongation of break (EB) of bionanocomposite samples.

Longer UVT also increased ME and decreased EB of the bionanocomposite (Figure 3b,c). The ME after 30 and 60 min UVT was increased by 107% and 254% compared to the untreated bionanocomposite due to good adhesion bonding between nanofiber cellulose and starch matrix [35,36]. UVT also compacts the structure [5]. This fact was supported by SEM observation of the fracture surface (Figure 2e,f). Previous research has reported that ultrasonic treatment reduces the porosities, improves the dispersion of fiber in matrix, the adhesion between matrix and fiber which results in high tensile strength [6]. The treated bionanocomposite in this research had a TS two times higher than water hyacinth microfiber reinforced TPS in a previous study [5], suggesting that nano-sized fibers and UVT treatment significantly improves the properties of bionanocomposites.

3.4. Crystallinity Index Analysis

The XRD formation of NFC water hyacinth, pure TPS, and all bionanocomposites is displayed in Figure 4. It can be seen that NFC shows two main peaks which indicates the cellulose type I. This is caused by interaction between hydroxyl groups of cellulose [21]. The NFC water hyacinth had a CI of about 80%. This value was higher than other sources such as water hyacinth processed by high shear homogenization [21], cassava bagasse cellulose nanofibrils [30], and garlic stalks [32].

Pure TPS film had two peaks around 2θ = 15–20°. The peak at 2θ around 17° indicates the amylopectin recrystallization (B-type crystallization). Meanwhile, the peak intensity at 2θ around 19.8° indicates VH-type recrystallization of amylose during cooling after fabrication because of the presence of glycerol, isopropanol, and lysophospholipids content [30]. According to the previous report, the calculation of CI used the Hulleman's method [27]. The CI of pure TPS film was 23%.

The addition of NFC in TPS matrix in a UVT treatment makes some changes in peak intensity. It can be seen that with longer UVT, peak intensity increases. This is probably due to a number of factors such as more complete gelation of starch during fabrication, homogeneity of the distribution of fiber in the starch matrix, and good adhesion bonding between fiber and matrix [6,37]. According to a previous report, amylose recrystallizes during gelation which changes the matrix structure and affects XRD peak intensity [6]. In addition, reduction of the agglomeration of fiber in the starch matrix will significantly influence the visible peak intensity [5,6].

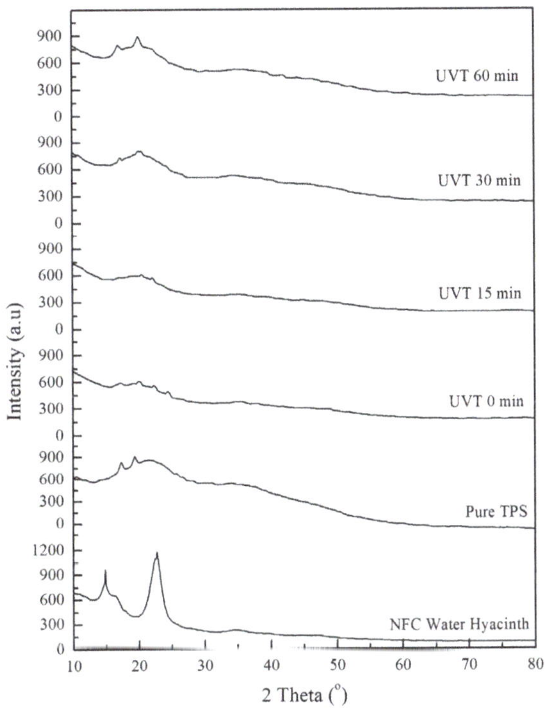

Figure 4. XRD pattern of all samples tested.

CI of TPS bionanocomposite after UVT of 0, 15, 30, and 60 min was 7, 8, 18, and 35%, respectively. Longer UVT leads to increasing CI. Despite the lower fiber loading, the UVT treated CI are comparable to CI of non-vibrated bionanocomposites from several other sources [24,30] and are significantly higher than sugarcane bagasse nanofiber (20 wt %) reinforced potato starch (24.4% CI) [24]. High CI corresponds to good mechanical properties (see Figure 3a).

4. Conclusions

UVT of yam bean TPS reinforced with a 1 wt % loading of water hyacinth NFC improved its mechanical properties. A 1 wt % loading of water hyacinth NFC reinforcement improved the mechanical properties of UVT of yam bean TPS. The maximum value tensile strength obtained was 11.4 MPa after 60 min UVT. The highest crystallinity index was 35% after 60 min UVT. Values obtained were in the same range as non-sonicated biocomposites from similar materials but with much higher nanofiber loadings. The improvement of mechanical properties and crystallinity index due to ultrasonication could enable a significant reduction of the amount of filler loading required in a matrix.

Author Contributions: M.A. conducted the experiments, wrote the paper and analyzed the data; H.A. as the supervisor of this research provides the addition of in-depth addition analyzed the data, improved the paper quality; A.K. and A.P. as the co-promotor provide additional explanation of characterization, M.M. and F.H. gave their energy and time to help the fabrication of bionanocomposite samples during experiment.

Acknowledgments: This research was funded by Directorate General of Higher Education Ministry of National Education (KEMENRISTEK DIKTI), Indonesia, with project name The Research of Master Program Leading to Doctoral Degree for Excellent Students (PMDSU Batch-2) in the year of 2018. We also thank Fay Farley for her proofreading.

Conflicts of Interest: The authors declare no conflict of interest.

References

1. Ahmed, T.; Shahid, M.; Azeem, F.; Rasul, I.; Shah, A.A.; Noman, M.; Hameed, A.; Manzoor, N.; Manzoor, I.; Muhammad, S. Biodegradation of plastics: Current scenario and future prospects for environmental safety. *Environ. Sci. Pollut. Res.* **2018**, *25*, 1–12. [CrossRef] [PubMed]
2. Shit, S.C.; Shah, P.M. Edible polymers: Challenges and opportunities. *J. Polym.* **2014**, *2014*, 427259. [CrossRef]
3. Kaewtatip, K.; Thongmee, J. Studies on the structure and properties of thermoplastic starch/luffa fiber composites. *Mater. Des.* **2012**, *40*, 314–318. [CrossRef]
4. Karimi, S.; Abdulkhani, A.; Tahir, P.M.; Dufresne, A. Effect of cellulosic fiber scale on linear and non-linear mechanical performance of starch-based composites. *Int. J. Biol. Macromol.* **2016**, *91*, 1040–1044. [CrossRef] [PubMed]
5. Asrofi, M.; Abral, H.; Putra, Y.K.; Sapuan, S.M.; Kim, H.J. Effect of duration of sonication during gelatinization on properties of tapioca starch water hyacinth fiber biocomposite. *Int. J. Biol. Macromol.* **2018**, *108*, 167–176. [CrossRef] [PubMed]
6. Abral, H.; Putra, G.J.; Asrofi, M.; Park, J.W.; Kim, H.J. Effect of vibration duration of high ultrasound applied to bio-composite while gelatinized on its properties. *Ultrason. Sonochem.* **2018**, *40*, 697–702. [CrossRef] [PubMed]
7. Mohanty, A.K.; Misra, M.; Drzal, L.T. Surface modifications of natural fibres and peformance of the resulting biocomposite: An overview. *Compos. Interfaces* **2001**, *8*, 313–343. [CrossRef]
8. Mali, S.; Grossmann, E.M.V.; Garcia, M.A.; Martino, M.N.; Zaritzky, N.E. Mechanical and thermal properties of yam starch films. *Food Hydrocoll.* **2005**, *19*, 157–164. [CrossRef]
9. Abral, H.; Lawrensius, V.; Handayani, D.; Sugiarti, E. Preparation of nano-sized particles from bacterial cellulose using ultrasonication and their characterization. *Carbohydr. Polym.* **2018**, *191*, 161–167. [CrossRef] [PubMed]
10. Lim, C.T. Nanofiber technology: Current status and emerging developments. *Prog. Polym. Sci.* **2017**, *70*, 1–17.
11. Asrofi, M.; Abral, H.; Kasim, A.; Pratoto, A. Characterization of the microfibrillated cellulose from water hyacinth pulp after alkali treatment and wet blending. *IOP Conf. Ser. Mater. Sci. Eng.* **2017**, *204*, 012018. [CrossRef]
12. Wahono, S.; Irwan, A.; Syafri, E.; Asrofi, M. Preparation and characterization of ramie cellulose nanofibers /CaCO3 unsaturated polyester resin composites. *ARPN J. Eng. Appl. Sci.* **2018**, *13*, 746–751.
13. Syafri, E.; Kasim, A.; Abral, H.; Asben, A. Effect of precipitated calcium carbonate on physical, mechanical and thermal properties of cassava starch bioplastic composites. *Int. J. Adv. Sci. Eng. Inf. Technol.* **2017**, *7*, 1950–1956. [CrossRef]
14. Fabra, M.J.; Martínez-Sanz, M.; Gómez-Mascaraque, L.G.; Gavara, R.; López-Rubio, A. Structural and physicochemical characterization of thermoplastic corn starch films containing microalgae. *Carbohydr. Polym.* **2018**, *186*, 184–191. [CrossRef] [PubMed]
15. Salaberria, A.M.; Labidi, J.; Fernandes, S.C.M. Chitin nanocrystals and nanofibers as nano-sized fillers into thermoplastic starch-based biocomposites processed by melt-mixing. *Chem. Eng. J.* **2014**, *256*, 356–364. [CrossRef]
16. Abdul Khalil, H.P.S.; Tye, Y.Y.; Saurabh, C.K.; Leh, C.P.; Lai, T.K.; Chong, E.W.N.; Nurul Fazita, M.R.; Mohd Hafiidz, J.; Banerjee, A.; Syakir, M.I. Biodegradable polymer films from seaweed polysaccharides: A review on cellulose as a reinforcement material. *Express Polym. Lett.* **2017**, *11*, 244–265. [CrossRef]

17. De Campos, A.; De Sena Neto, A.R.; Rondrigues, V.B.; Luchesi, B.R.; Moreira, F.K.V.; Correa, A.C.; Mattoso, L.H.C.; Marconcini, J.M. Bionanocomposites produced from cassava starch and oil palm mesocarp cellulose nanowhiskers. *Carbohydr. Polym.* **2017**, *175*, 330–336. [CrossRef] [PubMed]
18. Julie Chandra, C.S.; George, N.; Narayanankutty, S.K. Isolation and characterization of cellulose nanofibrils from arecanut husk fibre. *Carbohydr. Polym.* **2016**, *142*, 158–166.
19. Abral, H.; Anugrah, A.S.; Hafizulhaq, F.; Handayani, D.; Sugiarti, E.; Muslimin, A.N. Effect of nanofibers fraction on properties of the starch based biocomposite prepared in various ultrasonic powers. *Int. J. Biol. Macromol.* **2018**, *116*, 1214–1221. [CrossRef] [PubMed]
20. Cheng, W.; Chen, J.; Liu, D.; Ye, X.; Ke, F. Impact of ultrasonic treatment on properties of starch film-forming dispersion and the resulting films. *Carbohydr. Polym.* **2010**, *81*, 707–711. [CrossRef]
21. Asrofi, M.; Abral, H.; Kasim, A.; Pratoto, A. XRD and FTIR studies of nanocrystalline cellulose from water hyacinth (*Eichornia crassipes*) fiber. *J. Metastable Nanocryst. Mater.* **2017**, *29*, 9–16. [CrossRef]
22. Abral, H.; Dalimunthe, M.H.; Hartono, J.; Efendi, R.P.; Asrofi, M.; Sugiarti, E.; Sapuan, S.M.; Park, J.W.; Kim, H.J. Characterization of tapioca starch biopolymer composites reinforced with micro scale water hyacinth fibers. *Starch-Stärke* **2018**. [CrossRef]
23. Komuraiah, A.; Kumar, N.S.; Prasad, B.D. Chemical composition of natural fibers and its influence on their mechanical properties. *Mech. Compos. Mater.* **2014**, *50*, 359–376. [CrossRef]
24. Gilfillan, W.N.; Moghaddam, L.; Doherty, W.O.S. Preparation and characterization of composites from starch with sugarcane bagasse nanofibres. *Cellulose* **2014**, *21*, 2695–2712. [CrossRef]
25. Tharanathan, R.N. Biodegradable films and composite coatings: Past, present and future. *Trends Food Sci. Technol.* **2003**, *14*, 71–78. [CrossRef]
26. Segal, L.; Creely, J.J.; Martin, A.E.; Conrad, C.M. An empirical method for estimating the degree of crystallinity of native cellulose using the X-ray Diffractometer. *Text. Res. J.* **1959**, *29*, 786–794. [CrossRef]
27. Hulleman, S.H.D.; Kalisvaart, M.G.; Janssen, F.H.P.; Feil, H.; Vliegenthart, J.F.G. Origins of B-type crystallinity in glycerol-plasticised, compression-moulded potato starches. *Carbohydr. Polym.* **1999**, *39*, 351–360. [CrossRef]
28. Abraham, E.; Deepa, B.; Pothan, L.A.; Jacob, M.; Thomas, S.; Cvelbar, U.; Anandjiwala, R. Extraction of nanocellulose fibrils from lignocellulosic fibres: A novel approach. *Carbohydr. Polym.* **2011**, *86*, 1468–1475. [CrossRef]
29. Cherian, B.M.; Leao, A.L.; De Souza, S.F.; Thomas, S.; Photan, L.A.; Kottaisamy, M. Isolation of nanocellulose from pineapple leaf fibres by steam explosion. *Carbohydr. Polym.* **2010**, *81*, 720–725. [CrossRef]
30. Teixeira, E.D.M.; Pasquini, D.; Curvelo, A.A.S.; Corradini, E.; Belgacem, M.N.; Dufresne, A. Cassava bagasse cellulose nanofibrils reinforced thermoplastic cassava starch. *Carbohydr. Polym.* **2009**, *78*, 422–431. [CrossRef]
31. Mahardika, M.; Abral, H.; Kasim, A.; Arief, S.; Asrofi, M. Production of nanocellulose from pineapple leaf fibers via high-shear homogenization and ultrasonication. *Fibers* **2018**, *6*, 28. [CrossRef]
32. Agustin, M.B.; Ahmmad, B.; De Leon, E.R.P.; Buenaobra, J.L.; Salazar, J.R.; Hirose, F. Starch-based biocomposite films reinforced with cellulose nanocrystals from garlic stalks. *Polym. Compos.* **2013**, *34*, 1325–1332. [CrossRef]
33. Kargarzadeh, H.; Johar, N.; Ahmad, I. Starch biocomposite film reinforced by multiscale rice husk fiber. *Compos. Sci. Technol.* **2017**, *151*, 147–155. [CrossRef]
34. Hermansyah, H.; Carissa, R.; Faiz, M.B.; Deni, P. Food grade bioplastic based on corn starch with banana pseudostem fibre/bacterial cellulose hybrid filler. *Adv. Mater. Res.* **2014**, *997*, 158–168. [CrossRef]
35. Carmona-García, R.; Bello-Pérez, L.A.; Aguirre-Cruz, A.; Aparicio-Saguilán, A.; Hernández-Torres, J.; Alvarez-Ramirez, J. Effect of ultrasonic treatment on the morphological, physicochemical, functional, and rheological properties of starches with different granule size. *Starch-Stärke* **2016**, *68*, 972–979. [CrossRef]
36. Nascimento, P.; Marim, R.; Carvalho, G.; Mali, S. Nanocellulose produced from rice hulls and its effect on the properties of biodegradable starch films. *Mater. Res.* **2016**, *19*, 167–174. [CrossRef]
37. Monroy, Y.; Rivero, S.; García, M.A. Microstructural and techno-functional properties of cassava starch modified by ultrasound. *Ultrason. Sonochem.* **2018**, *42*, 795–804. [CrossRef] [PubMed]

fibers

MDPI

Article

Moisture Absorption and Opacity of Starch-Based Biocomposites Reinforced with Cellulose Fiber from Bengkoang

Fadli Hafizulhaq [1], Hairul Abral [1,*], Anwar Kasim [2], Syukri Arief [3] and Jon Affi [1]

[1] Department of Mechanical Engineering, Andalas University, Padang 25163, Sumatera Barat, Indonesia; hafizulhaq.fadli@gmail.com (F.H.); jonaffi@gmail.com (J.A.)
[2] Department of Agricultural Technology, Andalas University, Padang 25163, Sumatera Barat, Indonesia; anwar_ks@yahoo.com
[3] Department of Chemistry, Andalas University, Padang 25163, Sumatera Barat, Indonesia; syukriarief@gmail.com
* Correspondence: abral@ft.unand.ac.id; Tel.: +62-812-672-0355

Received: 17 July 2018; Accepted: 24 August 2018; Published: 29 August 2018

check for updates

Abstract: Cellulose fiber was isolated from bengkoang (*Pachyrhizus erosus*) tuber peel. A suspension consisting of distilled water, starch, and glycerol was mixed with various cellulose loadings (0, 2, 6, and 10 g) then gelatinized using a hot plate with a magnetic stirrer. The biocomposite gel was sonicated using an ultrasonication probe (47.78 W/cm^2 for 4 min). Scanning electron microscopy (SEM) micrographs for the fracture surface of resulting biocomposite films displayed a rougher surface than starch film, indicating fiber dispersion in the matrix. The opacity and moisture resistance of biocomposite films increased with the addition of cellulose. The opacity was at a maximum value (243.05 AUnm) with 10 g fiber, which was 11.27% higher than the starch film without cellulose. Moisture absorption of this biocomposite was 16.79% lower than the starch film. Fourier transform infrared (FTIR) confirmed this more hydrophobic nature with lower transmittance at –OH stretching in the composite than the starch film. The addition of cellulose fiber into the matrix also increased the crystallinity index.

Keywords: biocomposite; starch; cellulose; ultrasonication; moisture absorption; opacity

1. Introduction

In recent years, the demand for the development of starch-based films has been high because of their biodegradability. Films produced from starch have applications in food packaging and agricultural and medical technologies. Starch is abundant and cheap, and provides good film formability [1]. However starch-based films have high water sensitivity, are brittle, and exhibit poor mechanical properties [2,3].

Natural fiber can be used to improve the properties of starch film. Research has shown that fibers from natural sources can potentially provide promising reinforcing materials. Natural fibers are easy to obtain, low cost, non-toxic, and have a high specific strength due to their low density [4,5]. In addition, natural fibers are environmentally friendly and require little energy in the composite fabrication process [6]. Many researchers have reported the improvement of starch-based biocomposites reinforced with natural fiber. They increase their oxygen barrier properties [7,8], water and moisture resistance [2,9,10], and mechanical properties [11–13].

Starch and biocomposite film properties also can be improved with ultrasonication. During the gelatinization process, ultrasonication helps to dissolve ghost particles and clumps of insoluble starch that weaken the resulting film's mechanical properties [14,15]. Ultrasonication aids with dispersion and break-up of agglomerations of fibers in the matrix [16]. Also, ultrasonication is used in the production of nano-sized cellulose molecules because it produces high shear energy that breaks up long cellulose into shorter segments [17–19]. A reduction of cellulose size increases the area of contact surface between reinforcement and matrix, which improves the mechanical and barrier properties of the composite [20].

In last decade, the biocomposite potential of fibers and starches from various sources have been explored. Cassava, corn, sago, and potato starch have been used in the matrix. Fibers from potato tuber cells [10], banana peels [21,22], mandarin [16], grapefruit [23], and pomelo [24] have been used as reinforcement. Many of these starch and fiber sources can be obtained from agricultural products and waste. Another agricultural crop that produces both starch and cellulose fiber is bengkoang (*Pachyrhizus erosus*). This plant grows abundantly in tropical and subtropical regions, including in Indonesia, Mexico, Philippines, China, Malaysia, and Singapore. Information about starch extracted from the bengkoang tuber is still limited, and there has been no work reported on the use of the cellulose from bengkoang tuber peel as a reinforcement material. In this present study, we investigated the effect of various cellulose fiber loadings from bengkoang tuber peel on the properties of a starch-based biocomposite. Particle size distribution, opacity, SEM, X-ray diffraction (XRD), FTIR, and moisture absorption of the starch and biocomposite film were characterized.

2. Materials and Methods

2.1. Materials

Native bengkoang was obtained from a local farm at Padang, West Sumatra, Indonesia, and the starch (13.27% amylose content) was extracted. Glycerol purchased from Brataco (Jakarta, Indonesia) was used as a plasticizer. Distilled water was used as the plasticizing agent.

2.2. Isolation of Cellulose Fiber from Bengkoang Tuber Peel

Cellulose from bengkoang tuber peel was isolated using the method described by Julie Chandra et al. [17] with following modifications. Fibers were cut into 6 mm lengths. Alkaline treatment was used to remove hemicellulose and lignin content with 5% sodium hydroxide (NaOH) solution, and the suspension was heated at 50 °C for 4 h. After that, the alkali was washed from the fibers, and they were treated with 3.5 M hydrochloric acid (HCl) to break up the amorphous region, and the microfibrils were aggregate at 50 °C for 34 h. Treated fibers were washed with distilled water several times until pH = 7, then ground to a pulp manually with a mortar and pestle. This pulp was treated again with 5% NaOH solution at 50 °C for 4 h to remove the remaining non-cellulose content, and then subjected to acid hydrolysis with 5 M hydrochloric acid (HCl) at 50 °C for 17 h. At the end of process, a 4:1 mixture of sodium chlorite ($NaClO_2$) and glacial acetic acid was used to bleach the fibers for 2 h at 60 °C. All chemical treatments were conducted using a hot plate magnetic stirrer with continuous stirring at 500 rpm. Distilled water was used to neutralize the fibers from acid. About 10 mL of bleached fiber was mixed with 140 mL distilled water and ultrasonicated with a 20 KHz ultrasonic cell crusher (Model SJIA-1200W, Ningbo Yinzhou Sjia Lab Equipment Co., Ltd., Ningbo, China) at 47.78 W/cm^2 for 120 min, the temperature being maintained under 40 °C.

2.3. Biocomposite Film Preparation

10 wt% starch, 4 g glycerol, and 100 mL distilled water were mixed in a 250 mL beaker (diameter 70 mm, IWAKI, Sumedang, Indonesia). A magnetic stirrer (Scilogex MS-H280-Pro, Scilogex LLC, Rocky Hill, CT, USA) was used at 500 rpm, while the suspension was heated to 80 °C. After the suspension was gelatinized, it was ultrasonicated for 4 min using a 20 KHz ultrasonic cell crusher (Model SJIA-1200W, Ningbo Yinzhou Sjia Lab Equipment Co., Ltd., Ningbo, China) at 47.78 W/cm^2. The suspension was then cast to a petri dish (15 cm diameter) and dried in a ventilated oven for 20 h at 50 °C. The resulting film was labelled as the starch film. Cellulose was added to the suspension at the

beginning of the similar processes, to fabricate composite films with different amounts of fiber loading (2, 6, and 10 g, equivalent to a dried cellulose weight of 0.16, 0.48, and 0.80 wt% respectively) and labelled as biocomposites (BC), followed by a numerical code indication of the amount of fiber solution added. All suspensions of biocomposite were poured into petri dishes and dried in a ventilated oven for 20 h at 50 °C. Table 1 shows the composition of starch and biocomposite films.

Table 1. Composition of starch and biocomposite films.

Sample	Starch (g)	Glycerol (g)	Distilled Water (g)	Cellulose Suspension (g)
Control	10	4	100	-
BC-2	10	4	98	2
BC-6	10	4	94	6
BC-10	10	4	90	10

2.4. Characterization

2.4.1. Particle Size Distribution

Particle size measurement using dynamic light scattering (Zetasizer ZS, Malvern, UK) was used for the determination of particle size distribution of bengkoang tuber peel fiber. A 173° detector angle with a HeNe 4 mW 633 nm laser was used. Particle size determination was repeated three times.

2.4.2. Film Opacity

Opacity of films was determined with a spectrophotometer (Shimadzu UV 1800, Shimadzu Corporation, Kyoto, Japan). Films were cut into 1 cm × 2.5 cm rectangles and fixed onto the inner side of a spectrophotometer cell. The absorbance spectrum between 400 and 800 nm was recorded. The opacity of films was determined as the area under spectrum according to ASTM D 1003-00 (Standard test method for haze and luminous transmittance of transparent plastics). The opacity determinations were repeated three times.

2.4.3. X-ray Diffraction (XRD)

A PANalyticalXpert Pro diffractometer with Cu Kα radiation at 40kV and 30 mA was used to record XRD diffractograms of films. The diffraction between 2θ = 10° to 40° was scanned. The crystallinity index (*CI*) of films was determined by calculating the ratio of crystalline area to the total area on XRD diffractograms [8]. *CI* for fiber was calculated with following equation [25]:

$$CI = \frac{I_{cry} - I_{am}}{I_{cry}}$$

where I_{cry} was the maximum intensity of crystalline peak at 2θ = 22–24°, and I_{am} was the minimum intensity of amorphous diffraction at 2θ = 18–20°.

2.4.4. Fourier Transform Infrared Spectroscopy

FTIR spectra of films were recorded with a FTIR spectrometer (Frontier, PerkinElmer, Waltham, MA, USA) within the wavenumber range of 4000–600 cm^{-1} at 4 cm^{-1} resolution.

2.4.5. Moisture Absorption

Moisture absorption of all films was determined using the method described by Abral et al. [9] with the following modifications. Films were cut to 2 cm × 1 cm size and dried in a ventilated oven at 50 °C for 24 h. Dried pieces were weighed to determine the initial mass, then they were placed in a covered box containing saturated NaCl solution (Relative Humidity (RH) 75%) at 25 °C ± 2 °C. Every 30 min, all pieces were weighed.

2.4.6. Scanning Electron Microscopy

The morphology of starch and biocomposite film fracture surfaces were analyzed using a scanning electron microscope (HITACHI SU-3500, Hitachi High-Technologies Corporation, Tokyo, Japan) at an accelerating voltage of 5 kV in a 5×10^{-4} Pa vacuum. All samples were coated with gold (Au).

3. Results and Discussion

3.1. Particle Size Distribution

Figure 1 shows the size distribution of the cellulose particles produced from bengkoang peel fiber. 24.3% of the cellulose particles were below 100 nm in length which is within the nano-size range, but the other 72.5% were in the 100–250 nm micro-sized range. The reduction of cellulose size started from the removal of the amorphous components from the cellulose after acid hydrolysis [26,27], and then it was reduced further with ultrasonication as the high shear forces and cavitation energy broke up the particles. It also broke apart the aggregates of cellulose, enabling for better dispersion in the aqueous suspension. Similar results have also been reported by Abral et al. [18], who isolated nanocellulose from bacterial cellulose, and Niu et al. [28], who isolated nanocellulose from microcrystalline cellulose.

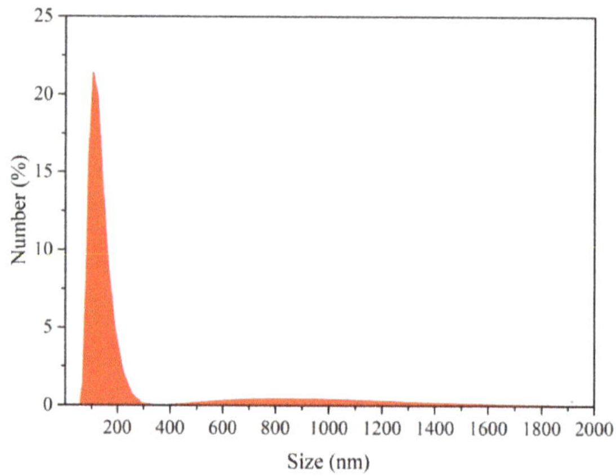

Figure 1. Particle size distribution of cellulose fiber from bengkoang tuber peel.

3.2. Film Opacity

Figure 2 shows the opacity of starch and biocomposite films. The addition of cellulose in the matrix increased the transparency of the biocomposites. For example, the opacity of the starch film was 218.44 AUnm, which was lower than that of BC-10 (243.05 AUnm, 11.27% higher). This is because cellulose is more opaque than starch film. Previous studies also reported on the increasing opacity of biocomposites with increasing fiber content [2,22].

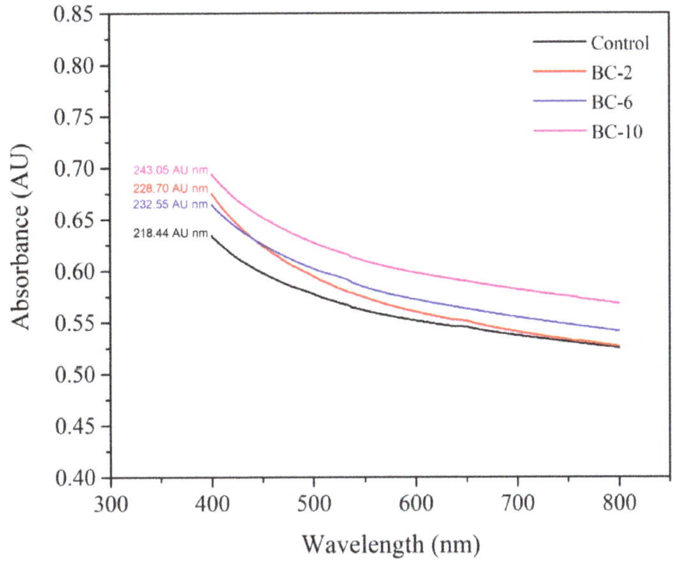

Figure 2. Film opacity of control and biocomposite films.

3.3. X-ray Diffraction

Figure 3 shows the XRD patterns of fibers, starch films, and biocomposite films. As shown in Figure 3b, two major peaks were observed at around 2θ = 17° and 22.2°. All films showed similar curves, but the diffraction intensity increased as the amount of cellulose increased from 1211 (Control) to 1351 (BC-10) at around 2θ = 22.2°. This result is because the presence of cellulose in the matrix *CI* for fiber was 58.48% higher than the starch and biocomposite films. In this case, crystallinity index (*CI*) for BC-10 was 2.51% higher than that for the control. A similar result was also reported by Abral et al. [13]; they reported the increasing *CI* of the composites with the addition of fibers.

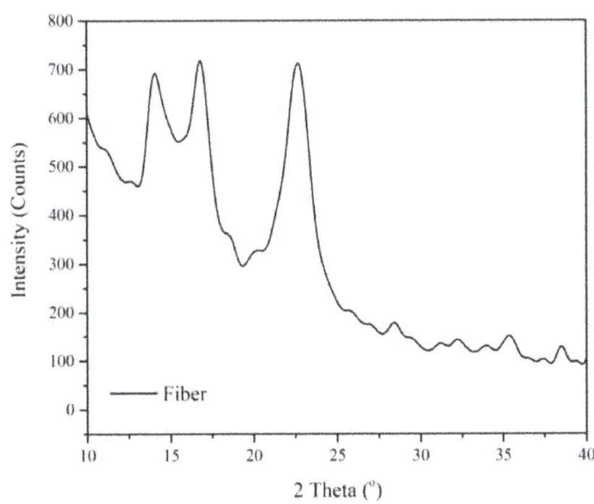

(a)

Figure 3. *Cont.*

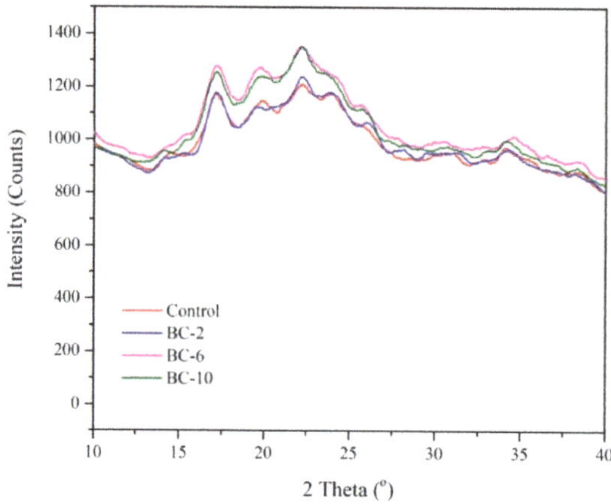

(b)

Figure 3. X-ray diffraction (XRD) patterns of (**a**) fibers, and (**b**) control and biocomposite films.

3.4. Fourier Transform Infrared

Figure 4 displays the FTIR spectra of starch and biocomposite films. All FTIR spectra showed similar patterns. Ultrasonication did not result in the appearance of new functional groups, but it shifted the transmittance intensity of the pattern. The band, at around 3296 cm^{-1} corresponded to the complex vibrational stretching of hydrogen bonded hydroxyl groups (–OH) [29]. Other bands were visible around 2926 cm^{-1} (–CH stretching), 1647 cm^{-1} (H–O–H stretching, absorbed water), 1353 cm^{-1} (–CO stretching in an aryl-arkyl ether) and 1013 (–CO stretching in an anhydroglucose ring). Transmittance intensity of all bands increased with an increase in fiber loading.

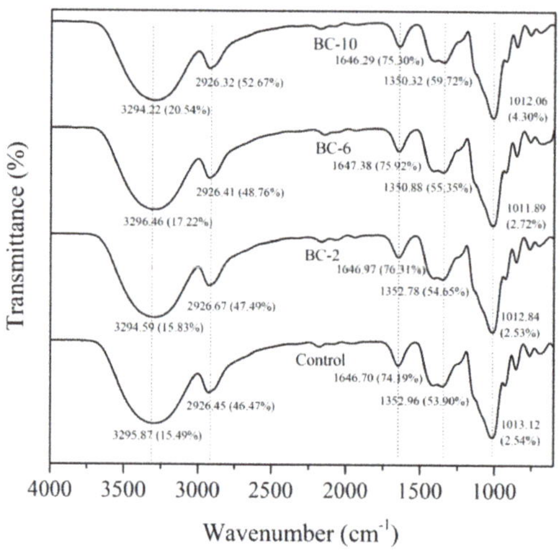

Figure 4. FTIR spectra of control and biocomposite films.

FTIR peaks can be also used to determine the change in the crystalline structure of films. This is assessed as the absorbance ratio of the peaks at 995/1022, this being the ratio between the crystalline and the amorphous regions of the biopolymer [14]. As shown in Table 2, the ratio was increased, as the fiber loading increased. For example, the ratio for BC-10 (0.59) was 22.92% higher than BC-2 (0.48). This is consistent with the crystallinity index calculated from the XRD pattern.

Table 2. Crystallinity index of starch and biocomposite films.

Sample	Crystallinity Index (%)	Intensity Ratio at 1022/995
Fiber	58.48	-
Control	44.70	0.47
BC-2	44.75	0.48
BC-6	45.63	0.54
BC-10	45.82	0.59

3.5. Moisture Absorption

Figure 5 shows the moisture absorption of starch and biocomposite films. The moisture absorption of the starch film was higher than that of the biocomposites. This is because the starch film was more hydrophilic than the biocomposites as reflected in the lower transmittance of –OH stretching bands than in the biocomposite. Adding 2 g fiber decreased the moisture absorption after 5 h by 3.85%. With further addition of the cellulose fiber, the moisture absorption continued to decrease. For example, the moisture absorption of BC-10 was 13.46% lower than that of BC-2. This was because the number of accessible hydroxyl groups in BC-10 was less than in BC-2. The addition of cellulose fiber reduced the total available –OH groups because of increased intermolecular bonding. This is consistent with the FTIR pattern (Figure 4) which confirmed the more hydrophobic nature of biocomposites with higher cellulose fiber loading in the matrix. For example transmittance (T) of –OH stretching for BC-2 was 2.20% higher than in the control. T value increased with increasing fiber loading, for instance BC-10 had a T value 29.75%, higher than that of BC-2. This result confirmed that BC-10 was more hydrophobic than BC-2. Moisture absorption of biocomposite film decreased as increasing fiber loading. Similar result have also been reported in a previous study which showed increasing moisture resistance of a biocomposite reinforced by higher fiber loading [13]. This is also consistent with the XRD curve (Figure 3). Similar agreement was also reported by Dufresne et al. [10], Montero et al. [20] and Abral et al. [30].

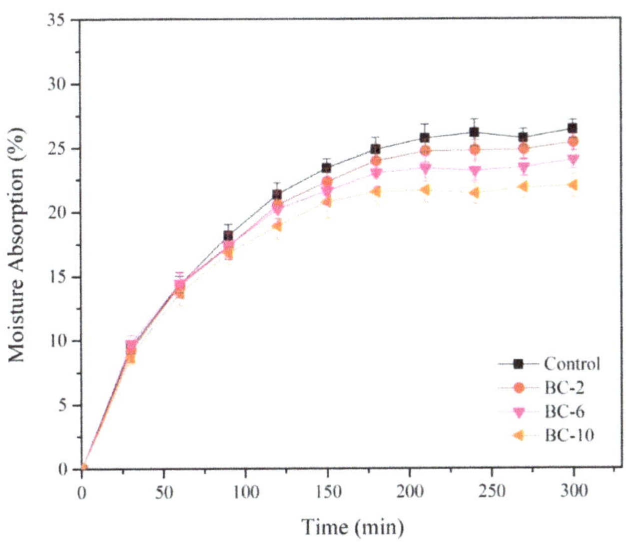

Figure 5. Moisture absorption for control and biocomposite films for 5 h testing time.

3.6. Scaning Electron Microscopy

Figure 6a–h show the SEM micrographs of the cross-sections of starch and biocomposite films. As can be seen in Figure 6c,d, the biocomposite film fracture surfaces were slightly rougher compared to the starch films (Figure 6a,b). Increases in cellulose suspension loading resulted in an increased roughness of fracture surfaces (Figure 6e–h). There is no agglomeration of cellulose fibers observed on the surface. This indicates that fibers were well dispersed in the starch matrix, and it may indicate increases in interfacial hydrogen bonding and homogeneity of the biocomposite structure. This result is in a good agreement with the previous study [13]. The higher compactness between the matrix and fiber led to an increase in the crystallinity index as shown in Table 2. The better interfacial hydrogen bonding reduced the number of accessible -OH groups, thus increasing moisture resistance of the biocomposites. This result is supported by the moisture absorption data in Figure 5, which shows a decrease in moisture absorption with increased fiber loading.

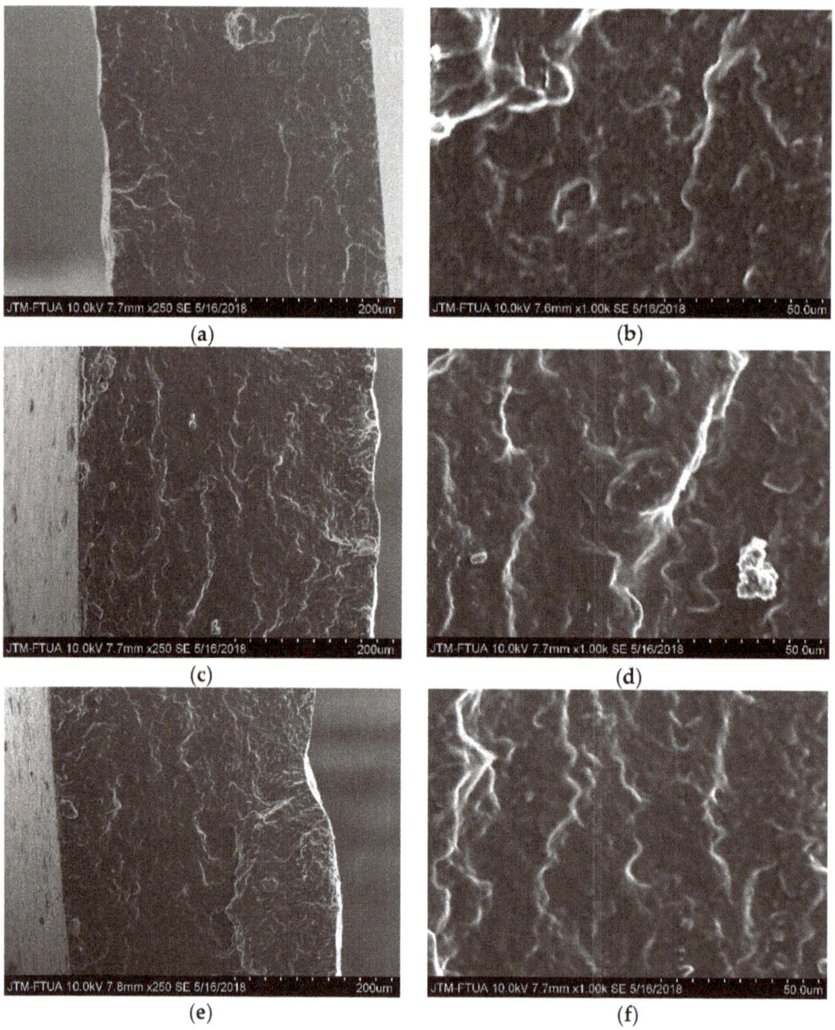

Figure 6. *Cont.*

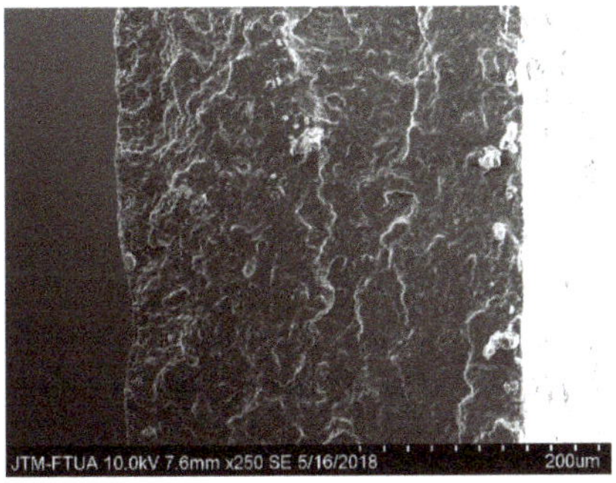

(g)

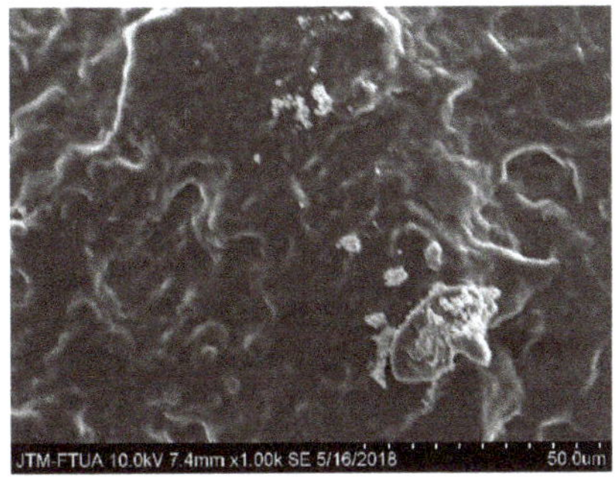

(h)

Figure 6. SEM micrograph of fractured cross section of the (**a**) Control; (**b**) BC-2; (**c**) BC-6; (**d**) BC-10 in X250; (**e**) Control; (**f**) BC-2; (**g**) BC-6; (**h**) BC-10 in X1000.

4. Conclusions

This study investigated the properties of starch-based biocomposites reinforced with cellulose fiber isolated from bengkoang tuber peel. Overall, cellulose fiber improves the opacity and moisture absorption of the starch-based film and within the range of cellulose fiber concentrations tested (up to 0.80 wt% for each 10 wt% of starch), the higher the cellulose fiber, the higher the opacity and moisture resistance. While moisture resistance is obviously important for the resilience of biofilm packaging, opacity is also an important parameter where the biofilms are used to contain items that need to be protected from light. The addition of cellulose fibers also increased the crystallinity index of the films. A high crystallinity index, which is related to the close and regular packing of polymers and microfibers, is likely to be an indication of superior mechanical properties. This indicates that bengkoang tuber peel fibers that were purified and reduced to micro-size in this manner have potential as a reinforcement material in biocomposite applications.

Author Contributions: Conceptualization, F.H. and H.A.; Methodology, F.H..; Validation, H.A.; Formal Analysis, F.H.; Investigation, F.H.; Data Curation, F.H., H.A. and J.A.; Writing—Original Draft Preparation, F.H.; Writing—Review & Editing, H.A.; Supervision, H.A., A.K. and S.A.

Funding: This research received no external funding.

Acknowledgments: This research was funded by the Directorate General of Higher Education, Ministry of Research, Technology, and Higher Education, Indonesia for supporting research funding, with the project name "Skim PMDSU", number 13/H.16/PMDSU/LPPM/2016.

Conflicts of Interest: The authors declare no conflict of interest.

References

1. Wilhelm, H.M.; Sierakowski, M.R.; Souza, G.P.; Wypych, F. Starch films reinforced with mineral clay. *Carbohydr. Polym.* **2003**, *52*, 101–110. [CrossRef]
2. Bodirlau, R.; Teaca, C.A.; Spiridon, I. Influence of natural fillers on the properties of starch-based biocomposite films. *Compos. Part B Eng.* **2013**, *44*, 575–583. [CrossRef]
3. Shah, U.; Naqash, F.; Gani, A.; Masoodi, F.A. Art and Science behind Modified Starch Edible Films and Coatings: A Review. *Compr. Rev. Food Sci. Food Saf.* **2016**, *15*, 568–580. [CrossRef]
4. Bledzki, A.K.; Gassan, J. Composites reinforced with cellulose based fibres. *Prog. Polym. Sci.* **1999**, *24*, 221–274. [CrossRef]
5. Saheb, N.; Jog, J. Natural Fiber Polymer Composites: A Review. *Adv. Polym. Technol.* **2015**, *2329*, 351–363. [CrossRef]
6. Rajesh, M.; Pitchaimani, J. Mechanical Properties of Natural Fiber Braided Yarn Woven Composite: Comparison with Conventional Yarn Woven Composite. *J. Bionic Eng.* **2017**, *14*, 141–150. [CrossRef]

7. Wang, J.; Gardner, D.J.; Stark, N.M.; Bousfield, D.W.; Tajvidi, M.; Cai, Z. Moisture and oxygen barrier properties of cellulose nanomaterial-based films Moisture and oxygen barrier properties of cellulose nanomaterial-based films. *ACS Sustain. Chem. Eng.* **2017**, *6*, 49–70. [CrossRef]
8. González, K.; Retegi, A.; González, A.; Eceiza, A.; Gabilondo, N. Starch and cellulose nanocrystals together into thermoplastic starch bionanocomposites. *Carbohydr. Polym.* **2015**, *117*, 83–90. [CrossRef] [PubMed]
9. Abral, H.; Putra, G.J.; Asrofi, M.; Park, J.W.; Kim, H.J. Effect of vibration duration of high ultrasound applied to bio-composite while gelatinized on its properties. *Ultrason. Sonochem.* **2018**, *40*, 697–702. [CrossRef] [PubMed]
10. Dufresne, A.; Dupeyre, D.; Vignon, M.R. Cellulose Microfibrils from Potato Tuber Cells: Processing and Characterization of Starch—Cellulose Microfibril Composites. *J. Appl. Polym. Sci.* **1999**, *76*, 2080–2092. [CrossRef]
11. Kakroodi, A.R.; Cheng, S.; Sain, M.; Asiri, A. Mechanical, Thermal, and Morphological Properties of Nanocomposites Based on Polyvinyl Alcohol and Cellulose Nanofiber from Aloe vera Rind. *J. Nanomater.* **2014**, *2014*, 12–18.
12. Asrofi, M.; Abral, H.; Kurnia, Y.K.; Sapuan, S.M.; Kim, H.J. Effect of duration of sonication during gelatinization on properties of tapioca starch water hyacinth fiber biocomposite. *Int. J. Biol. Macromol.* **2018**, *108*, 167–176. [CrossRef] [PubMed]
13. Abral, H.; Dalimunthe, M.H.; Hartono, J.; Efendi, R.P.; Asrofi, M.; Sugiarti, E.; Sapuan, S.M.; Park, J.W.; Kim, H.J. Characterization of tapioca starch biopolymer composites reinforced with micro scale water hyacinth fibers. *Starch/Staerke* **2018**, *70*. [CrossRef]
14. Garcia-Hernandez, A.; Vernon-Carter, E.J.; Alvarez-Ramirez, J. Impact of ghosts on the mechanical, optical, and barrier properties of corn starch films. *Starch/Staerke* **2017**, *69*, 1600308. [CrossRef]
15. Iida, Y.; Tuziuti, T.; Yasui, K.; Towata, A.; Kozuka, T. Control of viscosity in starch and polysaccharide solutions with ultrasound after gelatinization. *Innov. Food Sci. Emerg. Technol.* **2008**, *9*, 140–146. [CrossRef]
16. Hiasa, S.; Iwamoto, S.; Endo, T.; Edashige, Y. Isolation of cellulose nanofibrils from mandarin (*Citrus unshiu*) peel waste. *Ind. Crop. Prod.* **2014**, *62*, 280–285. [CrossRef]
17. Julie Chandra, C.S.; George, N.; Narayanankutty, S.K. Isolation and characterization of cellulose nanofibrils from arecanut husk fibre. *Carbohydr. Polym.* **2016**, *142*, 158–166.
18. Abral, H.; Lawrensius, V.; Handayani, D.; Sugiarti, E. Preparation of nano-sized particles from bacterial cellulose using ultrasonication and their characterization. *Carbohydr. Polym.* **2018**, *191*, 161–167. [CrossRef] [PubMed]
19. Chen, W.; Yu, H.; Liu, Y.; Chen, P.; Zhang, M.; Hai, Y. Individualization of cellulose nanofibers from wood using high-intensity ultrasonication combined with chemical pretreatments. *Carbohydr. Polym.* **2011**, *83*, 1804–1811. [CrossRef]
20. Montero, B.; Rico, M.; Rodríguez-llamazares, S.; Barral, L.; Bouza, R. Effect of nanocellulose as a filler on biodegradable thermoplastic starch films from tuber, cereal and legume. *Carbohydr. Polym.* **2016**, *157*, 1094–1104. [CrossRef] [PubMed]
21. Tibolla, H.; Pelissari, F.M.; Martins, J.T.; Vicente, A.A.; Menegalli, F.C. Cellulose Nanofibers Produced from Banana Peel by Chemical and Mechanical Treatments: Characterization and Cytotoxicity Assessment. *Food Hydrocoll.* **2017**, *75*, 192–201. [CrossRef]
22. Pelissari, F.M.; Andrade-mahecha, M.M.; José, P.; Sobral, A.; Menegalli, F.C. Nanocomposites based on Banana Starch Reinforced with Cellulose Nanofibers Isolated from Banana Peels. *J. Colloid Interface Sci.* **2017**, *505*, 154–167. [CrossRef] [PubMed]
23. Karataş, M.; Arslan, N. Flow behaviours of cellulose and carboxymethyl cellulose from grapefruit peel. *Food Hydrocoll.* **2016**, *58*, 235–245. [CrossRef]
24. Liu, Y.; Liu, A.; Ibrahim, S.A.; Yang, H.; Huang, W. Isolation and characterization of microcrystalline cellulose from pomelo peel. *Int. J. Biol. Macromol.* **2018**, *111*, 717–721. [CrossRef] [PubMed]
25. Segal, L.; Creely, J.J.; Martin, J.A.E.; Conrad, C.M. An Empirical Method for Estimating the Degree of Crystallinity of Native Cellulose Using the X-ray Diffractometer. *Text. Res. J.* **1959**, *29*, 786–794. [CrossRef]
26. Leite, A.L.M.P.; Florencia, Z.; Menegalli, C. Isolation and characterization of cellulose nanofibers from cassava root bagasse and peelings. *Carbohydr. Polym.* **2017**, *157*, 962–970. [CrossRef] [PubMed]

27. Khawas, P.; Deka, S.C. Isolation and characterization of cellulose nanofibers from culinary banana peel using high-intensity ultrasonication combined with chemical treatment. *Carbohydr. Polym.* **2016**, *137*, 608–616. [CrossRef] [PubMed]
28. Niu, F.; Li, M.; Huang, Q.; Zhang, X.; Pan, W.; Yang, J.; Li, J. The characteristic and dispersion stability of nanocellulose produced by mixed acid hydrolysis and ultrasonic assistance. *Carbohydr. Polym.* **2017**, *165*, 197–204. [CrossRef] [PubMed]
29. Sanjay, M.R.; Madhu, P.; Jawaid, M.; Senthamaraikannan, P.; Senthil, S.; Pradeep, S. Characterization and Properties of Natural Fiber Polymer Composites: A Comprehensive Review. *J. Clean. Prod.* **2017**, *172*, 566–581. [CrossRef]
30. Abral, H.; Anugrah, A.S.; Hafizulhaq, F.; Sugiarti, E.; Muslimin, A.N. Effect of nanofibers fraction on properties of the starch based biocomposite prepared in various ultrasonic power. *Int. J. Biol. Macromol.* **2018**, *116*, 1214–1221. [CrossRef] [PubMed]

fibers

MDPI

Article

Various Extraction Methods Influence the Adhesive Properties of Dried Distiller's Grains and Solubles, and Press Cakes of Pennycress (*Thlaspi arvense* L.) and Lesquerella [*Lesquerella fendleri* (A. Gary) S. Watson], in the Fabrication of Lignocellulosic Composites

Brent Tisserat [1,*], Fred Eller [1] and Rogers Harry-O'kuru [2]

1 Functional Foods Research Unit, National Center for Agricultural Utilization, Agricultural Research Service, U.S. Department of Agriculture, 1815 N. University Street, Peoria, IL 61601, USA; fred.eller@ars.usda.gov
2 Bio-Oils Research Unit, National Center for Agricultural Utilization, Agricultural Research Service, U.S. Department of Agriculture, 1815 N. University Street, Peoria, IL 61601, USA; rogers.harryokuru@ars.usda.gov
* Correspondence: brent.tisserat@ars.usda.gov; Tel.: +1-309-681-6289

Received: 22 February 2018; Accepted: 5 April 2018; Published: 24 April 2018

check for updates

Abstract: Lignocellulosic composite (LC) panels were fabricated using an adhesive matrix prepared from three different agricultural by-products: dried distillers grains with solubles (DDGS), pennycress (*Thlaspi arvense* L.) press cake (PPC), or lesquerella [*Lesquerella fendleri* (A. Gary) S. Watson] press cake (LPC) reinforced with *Paulownia elongata* L. wood (PW) particles. The goal in this study was to assess the mechanical properties of composites utilizing these low-cost matrix materials, which were subjected to various oil extraction methods. Three types of oil extraction methods were utilized: ethanol, supercritical CO_2, and hexane, in order to generate matrix materials. These matrix materials were mixed with equal proportions of PW and hot pressed to generate panels. Overall, hexane extraction was the best method to enhance the mechanical properties of the matrices used to fabricate lignocellulosic composites. LPC's produced a matrix that gave the resulting composite superior flexural properties compared to composites generated from DDGS and PPC matrices. The mechanical properties of composites generated from soy products (soybean meal flour or soy protein isolate) were similar to those derived from DDGS, PPC, or LPC. The dimensional stability properties of LCs were improved when the hexane extraction method was employed, unlike with the other extraction methods that were used to generate matrices.

Keywords: flexural properties; panels; by-products; non-dietetic uses; modulus of rupture; modulus of elasticity

1. Introduction

The economic viability of a biorefinery is increasingly dependent on the efficient disposal of the residual seed meals to obtain a profitable revenue source [1,2]. Grain ethanol dry milling operations generally sell their solid seed meal residue by-product, i.e., dried distillers grains with solubles (DDGS), as an animal feed [3–6]. In 2016, 44.5 million tons of DDGS was projected to be generated in the USA, selling for \$0.06–0.07/lb (\$0.13–0.15/kg) [7,8]. Health concerns have arisen concerning DDGS nutritional safety due to the presence of toxins and genetic modifications [3,4,7–12]. DDGS revenues significantly aid in the profitability of biorefineries; however, DDGS prices fluctuate based on the price of corn [1,13]. New product avenues to utilize DDGSs, which are not dependant on corn prices and nutritional health concerns, would benefit the ethanol industry.

Lesquerella [*Lesquerella fendleri* (A. Gary) S. Watson] and pennycress (*Thlaspi arvense* L.) are members of the family Brassicaceae and have been extensively studied as potential oil seed crops [2,14]. Lesquerella is native to southwestern USA, and its seeds contain ~30% oil and ~25% protein [2,15,16]. Lesquerella seeds contain high levels of the hydroxylated fatty acids, auricolic, and lesquereolic acids, which could compete with castor oil [2,14–16]. Pennycress is a winter annual grown in temperate North America and produces a seed that contains ~35% oil and ~30% protein [17,18]. The dominant fatty acid in pennycress is erucic acid, which could compete with rapeseed or canola oil [17]. However, neither plant is commercially harvested at this time due to their marginal profitability. In order to increase the revenues from these crops, other high-value products need to be identified from their seed meal residues [2,14]. Both plants have seed meals that contain high levels of glucosinolates, which precludes their use in human diets, as well as limits their use as an animal feed [18–20]. Two suggested non-dietary uses for pennycress press cake (PPC) are a soil biofumigant [20] or a filler/reinforcement material to be incorporated into plastic composites [21]. Non-dietary uses for lesquerella press cake (LPC) are a fertilizer [19], a tackifier adhesive [22,23], or a pre-cursor feed source to generate succinic acid via fermentation [24].

Engineered wood products are common building materials and are fabricated from sawmill scraps and wood wastes that are glued together by synthetic adhesives [25,26]. The majority of these adhesives is formaldehyde-based, which poses health and environmental problems [27–29]. As an alternative, bio-based adhesives using soybean meal flour (SBM), soy-protein hydrolysate, soy protein isolate (SPI), lignins, or tannins can partially or completely substitute for synthetic adhesives [27,30–32]. However, soy commercial products use as adhesives is minimal due to their high cost and higher value as a food additive [27]. Low cost and effective bio-based adhesives are needed by the engineered wood industry to replace synthetic adhesives. In this study, the adhesive properties of DDGS, LPC, and PPC in the fabrication of engineered panels were evaluated.

Our initial results revealed that non-defatted meals that retained residual oils were inferior adhesives compared to meals that were defatted [33]. This suggests that vegetable oil interferes with interfacial binding of the seed meal flour to the wood particles. Further, oil removal from meals increases protein concentration, which aids in adhesive bonding. Similarly, in other studies with plastic composites, DDGS and pennycress press cakes that were defatted prior to use made superior composites compared to those that were made with the non-defatted materials [21,34]. The focus of this study was to evaluate the mechanical properties of engineered panels fabricated with adhesive/resins derived from DDGSs and press cakes subjected to various solvent extraction techniques. Four distinct flours for each press cake or DDGS source were generated: no extraction/original material, ethanol extraction, hexane extraction, and supercritical CO_2 extraction. These methods of extraction are commonly used to extract vegetable oils from seed meal. Often, the industrial oil extraction process utilizes a mechanical extraction step whereby the oils are pressed out, producing a press cake or oilcake that contains some residual oil [35]. This press cake can be further treated with a solvent (typically *n*-hexane) to remove residual vegetable oils. Hexane solvent extraction is commonly used, but it has environmental and health issues [35]. Supercritical CO_2 and ethanol solvents are alternatives to hexane extraction [35–37]. Ethanol extraction has been found to remove sugars, oils, tocopherols, and phospholipids compared to using hexane extraction; fewer waxes were removed by ethanol extraction than by hexane extraction [35]. Supercritical CO_2 extraction utilizes the fluid state of CO_2 at or slightly above its critical temperature and critical pressure. Supercritical CO_2 is a common commercial industrial solvent because of its low toxicity and low environmental impact. The relatively low temperature employed in this process and the stability of CO_2 allows most compounds to be extracted with minimal denaturing of the original material [37]. Essentially, using the supercritical CO_2 solvent allows for removal of oils, which mimics hexane extraction.

Following the extraction step, original or defatted materials were then pulverized and sieved to a flour and employed as adhesives/resins in the preparation of high density fiberboard (HDF) panels via hot-pressing. In this study, a HDF panel was fabricated containing 50% adhesive and 50% *Paulownia elongata* L. wood (PW) family Paulowniaceae. No other additives were included in the composition that could detract or alter the adhesive nature of the matrix materials. The mechanical properties of these HDF panels were compared to the mechanical properties of panels made with soybean meal flour (SBM) or soybean protein isolate (SPI). SBM and SPI are typically employed as bio-based adhesives/resins in engineered wood composite studies [29–32]. The ultimate goal of this study was to identify the conditions that would generate alternative matrix adhesive materials that compare favorably to soy flour products. The development of technology to employ inexpensive and abundant DDGS and press cakes as an adhesive/resin matrix would eliminate the use of formaldehyde-containing substances in engineered wood and provide product avenues for these abundant and inexpensive agricultural by-products.

2. Materials and Methods

2.1. Materials and Processing Employed

DDGS were obtained from a corn dry milling processing plant located in Peoria IL (Archer Daniels Midland, Decatur, IL, USA). Lesquerella seeds were obtained from a crop grown in Arizona, and pennycress seeds were obtained from a crop grown in Peoria County, IL. Lesquerella and pennycress seeds were initially milled between rollers (Model SP900-12 roller miller, Roskamp Champion, Waterloo, IA, USA) with a gap set at 0.25 mm. Milled seeds were preheated in a cooker to 93 °C for 75 min then screw pressed with a laboratory screw press (Model L-250, French Oil Mill Machinery Co., Piqua, OH, USA) to remove crude oil and generate press cakes [38]. Press cakes (LPC and PPC) and DDGS were then treated with hexane extraction via Soxhlet extractor, supercritical CO_2 extraction, or ethanol extraction. Hexane extractions (H) were performed with a Soxhlet apparatus with the condenser attached to a recirculating chiller (Merlin™ Model M75 LR, ThermoFisher Scientific, Waltham, MA, USA). The 400 g of material was placed in a cellulose thimble (90 mm diam × 200 mm length × 1 mm thickness) (I.W. Tremont LabExact®, Hawthorne, NJ, USA). Extraction conditions were as follows: solvent temperature, 70 °C; condenser cooling temperature, 12 °C; solvent volume, 6 L; and extraction time 120 h. Supercritical CO_2 extractions (SC-CO_2) were performed in a modified Hewlett-Packard 7610A gas chromatograph oven (Hewlett-Packard, Avondale, PA, USA) with plumbing as described by Friedrich and List [39]. The material to be extracted was placed in a 100-mL stainless steel extraction vessel (Thar Technologies, Inc., Pittsburgh, PA, USA) with 30-mm diam Whatman glass microfiber filter circles (Fisher Scientific, Waltham, MA, USA) on each end. Extraction conditions were as follows: temperature, 80 °C; pressure, 55.2 MPa; the solvent to feed (S:F) ratio, 15:1 (g/g); and expanded CO_2 flow rate, ca. 3 L/min. A total of ca. 600 g of each material was extracted in 8–9 separate batches (i.e., individual batches of between 65 g and 80 g). Ethanol extractions (ETOH) were also performed in the modified gas chromatograph oven. The material for these extractions was placed in a 1-L stainless steel extraction vessel (Thar Technologies, Inc., Pittsburgh, PA, USA) with 7.62-cm diam Whatman paper filters (Fisher Scientific, Waltham, MA, USA) and glass microfiber filters on each end. Extraction conditions were as follows: temperature, 80 °C; pressure, 10.3 MPa; S:F ratio, 5:1 (g/g); and ethanol flow rate, ca. 10 mL/min. A total of ca. 700–800 g of each material was extracted.

After extractions, DDGS and press cake materials were ball ground into a flour (<74 μm particles) using a laboratory bench top ball mill (Model 801CVM, U.S. Stoneware, East Palestine, OH, USA). Materials were added to alumina mill jars containing grinding pellets (U.S. Stoneware) and ground at 50 rpm for 48 h. DDGS or press cake flours were sieved via a #80 screen and were composed of ≤250 μm particles. These were the matrix materials employed in this study.

Defatted Soybean meal (SBM) flour contained ~52% protein (Prolia (200/90, Cargill Inc., Cedar Rapids, IA, USA). Soy protein isolate (SPI) flour contained ~89% protein (PRO-Fam 974; ADM, Decatur, IL, USA). Soy materials were used as provided.

Wood reinforcements materials consisted of PW shavings procured from 3-year-old trees grown in Fort Valley, GA. Shavings were successively milled through 4-, 2-, and then 1-mm screens with a Thomas-Wiley mill grinder (Model 4, Thomas Scientific, Swedesboro, NJ, USA). Next, the particles were sized employing a Ro-Tap™ Shaker (Model RX-29, Tyler, Mentor, OH, USA) fitted with 203 mm diameter stainless steel screen/sieves of #10, #12, and #30 US Standards sizes (Newark Wire Cloth Company, Clifton, NJ, USA). Shaker was operated for 60 min intervals at 278 rpm. Two PW mixtures were used throughout this study composed of 50% ≤600 μm particle selection obtained from particles passed through the #30 mesh sieve and a >600–1700 μm particle fraction obtained from particles passed through the #12 mesh sieve, and were collected on the #30 mesh sieve. Moisture content of matrix materials was ~6–8%, and PW was ~8–9%.

2.2. Compositional Analysis Method

Crude protein content of untreated and extracted DDGS and press cake samples was derived through combustion using a Protein/Nitrogen Determinator (LECO FP-528 Model 601-500, St. Joseph, MI, USA). Moisture content of the untreated and extracted samples was determined using a Halogen Moisture balance/analyzer (Model HG63, Mettler-Toledo International Inc., Columbus, OH, USA).

2.3. Panel Preparations

Lignocellulosic panels were composed of 80 g of the adhesive matrix (i.e., DDGS, LPC, PPC, SBM, or SPI) respectively, mixed with 40 g of ≤600 μm PW particles and 40 g of 600–1700 μm PW particles (Table 1). Mixing of matrix and PW was performed in a zip-lock bag using circular agitation for 15 min in a compact dryer (Model MCSDRY1S, Magic Chef, Chicago, IL, USA). Materials were then transferred to an aluminum mold (outer dimensions: 15.2 cm W × 30.5 cm L × 5 cm D; mold cavity: 12.7 cm W × 28 cm L × 5 cm D). Pressings were conducted using manual hydraulic presses (Model 4126, Carver Press Inc., Wabash, IN, USA), preheated to 185 °C under 5.6 MPa pressure for 12 min, and then heating was discontinued and the platens cooled using cold water to room temperature while still under pressure.

Table 1. Weight percentage (wt. %) of tested composite formulations.

Formulation Code	Extraction Method [1]	Matrix (wt. %)	PW [2] (wt. %)
DDGS-PW	–	50	50
DDGS/E-PW	ETOH	50	50
DDGS/CO_2-PW	SC-CO_2	50	50
DDGS/H-PW	H	50	50
PPC-PW	–	50	50
PPC/E-PW	ETOH	50	50
PPC/CO_2-PW	SC-CO_2	50	50
PPC/H-PW	H	50	50
LPC-PW	–	50	50
LPC/E-PW	ETOH	50	50
LPC/CO_2-PW	SC-CO_2	50	50
LPC/H-PW	H	50	50
SPI-PW	–	50	50
SBM-PW	–	50	50

[1] Extraction method: ETOH = ethanol extraction; SC-CO_2 = supercritical CO_2 extraction; H = hexane extraction;
[2] PW fraction contains equal portions of ≤600 μm particles and 600–1700 μm particles.

2.4. Mechanical and Dimensional Property Measurements

Following pressing, composite panels were conditioned for 96 hs at 25 °C and 50% relative humidity. Specimen test board dimensions were then cut using a table saw as prescribed by EN

310:1993 procedures [40]. Panels were cut into test sample with dimensions of 50 mm W × 127 mm L × ≈ 3.5–5.5 mm thickness. The modulus of elasticity (MOE) and modulus of rupture (MOR) tests were conducted on five test samples for each formulation using a universal testing machine (Instron Model 1122 (Instron Corp., Norwood, MA, USA) with a crosshead speed of 5 mm/min equipped with a 1 kg load cell. Density of panels was measured using EN 323:1993 standard procedures, employing five samples for each formulation [41]. Following flexural analysis panels were photographed with a digital camera equipped with a 5× optical/2× digital zoom lense (Model # DSCF707 Cyber-shot 5 MP, Sony Corp., Tokyo, Japan). Surface and sawn cross sections of panels were examined and photographed.

Water absorption (WA) and thickness swelling (TS) of composite samples were conducted on samples cut into 50 mm^2 squares that were immersed into distilled water for 24 h following EN 317:1993 procedure [42].

For each composite formulation, five specimens were tested and their averages and standard errors reported. The experimental data was analyzed statistically by analysis of variance for statistical significance and multiple comparisons of means using Duncan's Multiple Range Test ($p \leq 0.05$).

3. Results and Discussion

3.1. Amino Acid, Moisture, Oil, and Protein Evaluation

As shown in Table 2 the protein, oils, and moisture content varied somewhat among the DDGS, LPC, and PPC matrices. The protein quality and quantity in various matrices is considered to be the major factor responsible for their adhesive properties [26–28,30–32]. For the DDGS matrix fractions, DDGS/E matrix exhibited the highest protein concentration (33.1%) while DDGS matrix had the lowest (26.6%). Since this study is interested in developing an adhesive product from these materials, we compared their protein contents to soybeans, which are the commercial standards for bio-adhesives. Protein adhesives (e.g., animal protein, caseins, and soy flours) have been employed to bond wood products for several years [30]. The DDGS and press cake materials have considerably less protein than soybean materials (Table 2). Prolia and SPI contain 52% and 89% protein, respectively. The final calculated amount of protein in the composite panels for the by-product matrices varied from 13.3 (DDGS) to 17.5% (PPC/E), while the composite panels employing SBM and SPI were calculated to contain 24.3% and 44.9%, respectively.

Table 2. Percent ingredients obtained from the chemical analysis of DDGS, press cakes, and soy products and the percent protein and lignocellulosic (LC)/panel.

Matrix	Protein (%)	Moisture (%)	Oil (%)	Other (%)	Protein/Panel (%)	LC/Panel (%)
DDGS	26.6	6.1	8.0	59.3	13.3	75.4
DDGS/E	33.1	5.8	n.d. [1]	61.1	16.6	76.3
DDGS/CO_2	28.5	5.9	n.d.	65.7	14.2	78.6
DDGS/H	31.2	5.8	n.d.	63.0	15.6	77.2
PPC	29.6	8.2	8.1	54.1	14.8	72.8
PPC/E	35.0	8.1	n.d.	56.9	17.5	74.2
PPC/CO_2	34.0	8.2	n.d.	57.9	17.0	74.7
PPC/H	30.1	8.2	n.d.	61.7	15.0	76.6
LPC	28.0	8.0	7.2	56.8	14.0	74.1
LPC/E	32.6	8.2	n.d.	59.2	16.3	75.4
LPC/CO_2	32.0	8.1	n.d.	59.9	16.0	75.7
LPC/H	31.3	8.2	n.d.	60.5	15.6	76.0
SBM	52.0	5.1	n.d.	44.4	24.3	69.0
SPI	89.0	3.2	0.0	6.8	44.9	49.3

[1] n.d. = not detected.

Moisture content in the ingredients is important, because excess water disrupts production speed and product quality [30]. Similarly, we found that employment of ingredients high in moisture

(25–30%) resulted in unsatisfactory panels that exhibited blistering and cracking. This was caused by excessive steam and water explusion during the molding process. Ingredients contained a relatively low moisture content of less than 9%. As shown in Table 2, the moisture content of DDGS, LPC, and PPC varied. The amino acid profiles of the various materials studied herein are presented in Table 3.

Table 3. Amino acid profiles expressed as percentage of dry matter of the crude protein for untreated DDGS, PPC, and LPC compared with SBM.

Amino Acid	Functional Group Characteristics	DDGS [43]	PPC [44]	LPC [45]	SBM [46]
Nonpolar:					
Alanine	hydrophobic	7.4	5.1	4.9	4.4
Isoleucine	hydrophobic	4.3	4.1	3.9	4.8
Leucine	hydrophobic	12.8	7.8	6.4	7.8
Methionine	hydrophobic	2.3	1.5	1.8	1.3
Phenylalanine	hydrophobic	5.5	4.8	4.6	5.0
Proline	hydrophobic	7.8	5.8	7.3	5.0
Trytophan	hydrophobic	0.8	1.3	0.9	1.0
Tyrosine	hydrophobic	4.7	3.3	3.1	3.6
Valine	hydrophobic	5.8	5.5	5.5	5.1
Total:		**51.4**	**39.2**	**38.5**	**37.9**
Polar:					
Arginine	positive charged/basic	5.5	7.7	8.0	7.3
Histidine	positive charged/basic	3.1	2.6	2.7	2.7
Lysine	positive charged/basic	3.9	5.4	6.6	6.6
Aspartic acid	negative charged/acidic	6.6	8.9	8.2	11.4
Glutamic acid	negative charged/acidic	12.8	17.1	15.3	17.2
Serine	uncharged	4.7	4.4	4.1	4.5
Threonine	uncharged	4.3	4.9	4.6	3.8
Total:		**40.9**	**51.0**	**49.5**	**53.5**
Other:					
Cystine	thiol	2.0	1.8	2.6	1.6
Glycine	hydrogen	4.3	6.9	6.6	4.3
Taurine	sulfonic acid	–	–	–	0.2
Hydroxy lysine		–	0.0	–	–
Hydroxyproline		0.4	1.1	2.8	–
Lanthionine		–	–	–	–
Ornithine		0.4	–	–	–
Total		**7.0**	**9.8**	**12.0**	**6.0**

The percentages of amino acid varied somewhat among the press cakes, DDGS, and SBM. The functional groups of amino acids of the proteins interact with reinforcement material (wood) and contribute toward adhesion [47]. Amino acids are often classified into general groups based on their polarity and charge [47,48]. DDGS were found to have a higher percentage of non-polar hydrophobic amino acids versus polar hydrophilic amino acids than PPC, LPC, or SBM. The individual amino acid profiles are given only for archival information, since we do not know how these amino acid functional groups are oriented in the protein types.

Essentially, there are four general protein types in plant seeds, and the specific type and quantity is distinct for each species (Table 4) [43–47]. Therefore, the bio-based adhesive composition is composed of a distinct protein profile (Table 4). For example, SBM consists of 80% globulins (glycinin and conglycinin) [30,47], while for DDGS 50% of the proteins were prolamins (specifically zein) [43]. The main protein types in PPC and LPC were albumins, globulins, and glutelins [44,45].

Table 4. Protein types in seed meals based on their percentage of crude protein.

Type	Extraction Solvent	DDGS [43]	PPC [44]	LPC [45]	SBM [47]
Albumin	Water	2	34	10	–
Globulin	0.5 M NaCl	8	13	24	80
Prolamin	70% ETOH	50	–	1	–
Glutelin	0.1 M NaOH	17	23	19	–
Other	Unextracted	24	30	45	20

Very little is actually known as to how these plant proteins manifest adhesive properties, although models have been proposed [30,47–52]. Seed proteins exist in quaternary structures that have little or no adhesive properties. Seed proteins are characterized as globular structures with a hydrophobic interior [48]. Denaturation of these proteins causes them to unfold into tertiary and secondary structures, allowing them to express interior amino acids. In this study, seed proteins subjected to high temperatures and pressure were able denature and express adhesive properties. As the proteins denature or unfold, the hydrophilic and hydrophobic groups of the amino acids interact with wood carbohydrate portion through hydrogen bonding, Van der Waal forces, and cross linking to create adhesion [47,48].

We attribute the majority of the extractables to residual oils. The type of extraction method affecting extraction yields (i.e., weight of material removed from the original seed meal) for each seed meal is shown in Table 5. The hexane extraction method (H) probably extracted primarily residual oils. When SC-CO_2 was employed, lesquerella exhibited the highest extraction yield, while DDGS exhibited the lowest yield. This suggests that lesquerella contains more non-polar material that is extracted by the non-polar SC-CO_2. However, when ETOH was employed, DDGS gave the highest yield, while pennycress gave the lowest yield. We can attribute this to the removal of the residual corn oil, proteins (zein), solubles, and carbohydrates from DDGS. Winkler et al. [53] similarly reported that ETOH extraction gave higher yields for DDGS than using hexane or SC-CO_2 extraction. DDGS apparently contain a higher amount of polar compounds that are removed by the relatively polar ethanol.

Table 5. Extraction yields (%) from DDGS, PPC, and LPC using various methods.

Extraction Method	DDGS	PPC	LPC
SC-CO_2	7.6%	9.5%	10.5%
ETOH	24.7%	17.4%	17.9%
H	8.0%	8.1%	7.2%

3.2. Adhesive Properties of Materials

The optical images of the flexural panels provide information concerning how the matrix and reinforcement materials interact (Figures 1 and 2). The surface portions of the panels had a smooth topography and did not exhibit any obvious imperfections. Visible differences between the composites were mainly those of color (Figures 1 and 2). The dark portions of the panels probably represent over-concentration of the matrix materials. Generally, composites containing the original untreated matrices were usually darker in color than the composites containing the treated matrices. The cross sections revealed color changes for the matrices. For example, the LPC-PW and LPC/H-PW composites were considerably darker than the LPC/E-PW and LPC/CO_2-PW composites (Figures 1 and 2). The sawn cross sections of composites also had a smooth topography (Figure 2). Wood particles were immersed in the matrices in a random manner. Clumping of wood particles was not observed, which indicates adequate distribution of the matrices and wood particles within the composites.

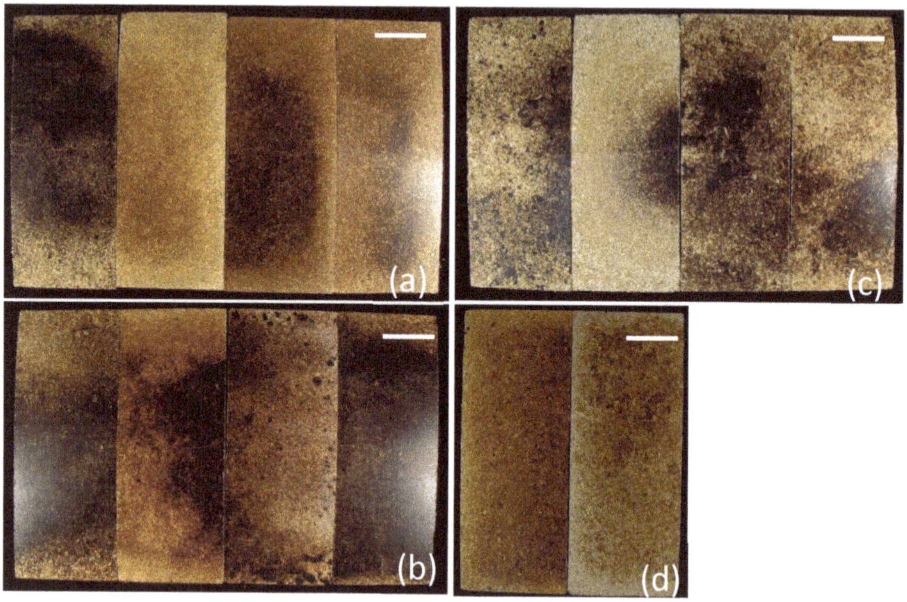

Figure 1. Examples of composite panels employed in flexural testing. (**a**) from left to right: DDGS-PW, DDGS/E-PW, DDGS/CO_2-PW, and DDGS/H-PW; (**b**) from left to right: LPC-PW, LPC/E-PW, LPC/CO_2-PW, and LPC/H-PW; (**c**) from left to right: PPC-PW, PPC/E-PW, PPC/CO_2-PW, and PPC/H-PW; (**d**) from left to right: SBM-PW and SPI-PW. Bar = 25 mm.

Figure 3 shows the flexural properties of panels fabricated in this study. The method of extraction employed on the matrix material had significant effects on the resulting panels' flexural properties. Generally, superior flexural properties were obtained from composites when the matrix material was subjected to extraction methods versus the untreated original matrix materials. For example, DDGS-PW panels exhibited modulus of rupture (MOR) and modulus of elasticity (MOE) values of 21.7 ± 1.9 MPa and 3365 ± 300 MPa, respectively, while DDGS/H-PW composite panels exhibited MOR and MOE values of 28.9 ± 2.2 MPa and 4965 ± 460 MPa, respectively. This translates in into percent increase in MOR and MOE values of +32% and +22%, respectively. For DDGS composites, no difference in flexural properties occurred comparing DDGS-PW to DDGS/E-PW, while significantly higher flexural properties were obtained from DDGS/CO_2-PW and DDGS/H-PW composites (Figure 3). In the case of DDGS/E-PW, the ethanol extraction appeared to be considerably less effective, probably due to the inadvertent removal of both carbohydrates and prolamin (zein) proteins. It is interesting to note that although this treatment (DDGS/E-PW) increased overall protein content, it still did not enhance the flexural properties over the untreated DDGS-PW composite. It is well documented that ethanol solvent is notable for removing prolamins in plant seeds [43–47]. The H and SC-CO_2 extraction generated a DDGS matrix material that had greater interfacial adhesion to the wood materials. We attribute this result to the effective removal of residual oils by these treatments. This situation is somewhat similar to that observed in hexane extraction of the DDGS results in a superior plastic composite when compared to plastic composites containing untreated DDGS [34].

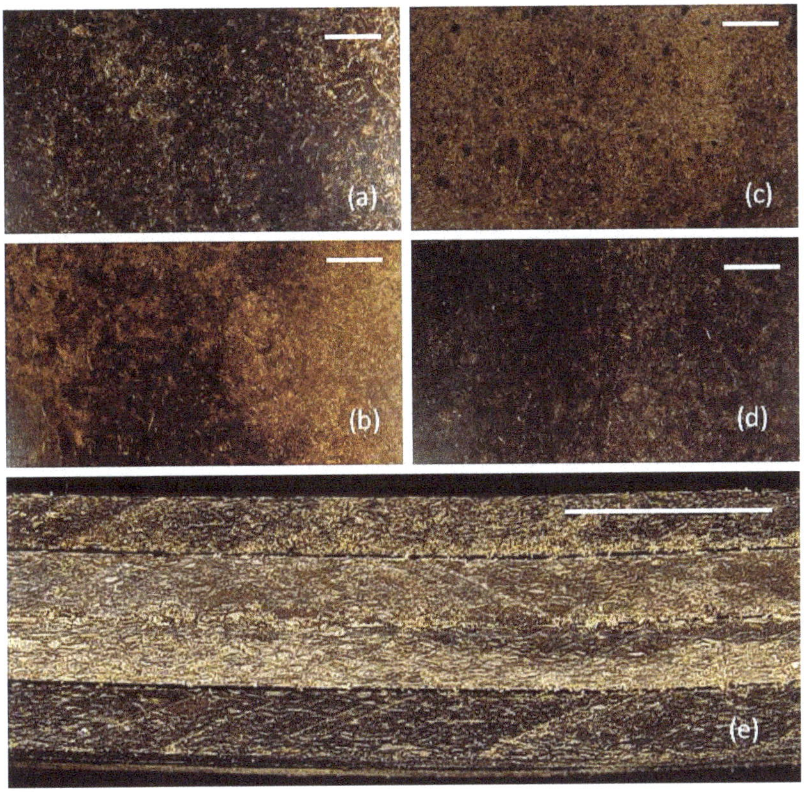

Figure 2. Optical images of surface and sawn cross sections of lesquerella composites. (**a**) LPC-PW surface; (**b**) LPC/E-PW surface; (**c**) LPC/CO_2-PW surface; (**d**) LPC/H-PW surface; and (**e**) cross sections of biocomposites, from top to bottom: LPC-PW, LPC/E-PW, LPC/CO_2-PW, and LPC/H PW. Bar = 10 mm.

For the pennycress composites, the untreated control (PPC-PW) exhibited the lowest flexural properties compared to the composites resulting from treated substrates (Figure 3). For example, the MOR and MOE of PPC/E-PW, PPC/CO_2-PW, and PPC/H-PW composites were +101% and +114%, +149% and +140%, and +155% and +148% greater, respectively, than the MOR and MOE values of PPC-PW composite. However, the PPC/E-PW composite exhibited lower flexural values compared to the flexural properties exhibited by PPC/CO_2-PW and PPC/H-PW composites. We attribute this situation to the type of extraction method employed. Reifschneider et al. [21] reported superior mechanical performance of PPC-plastic composites if the oils were removed by hexane compared to composites prepared with PPC containing oils. Clearly, removal of oils from the agriculture by-products improves the resultant composite's performance (Figure 3).

The highest MOE and MOR values were obtained from the LPC compared to the DDGS and PPC composites. For example, MOR and MOE values of LCP-PW were +85% and +163% greater than MOR and MOE values of DDGS-PW. This suggests that LCP may have some additional adhesive properties over those of the other two materials tested that not attributed to the protein content. This observed characteristic of the LPC matrix can be rationalized from earlier reported composition of the seed polysaccharide, namely, the seed surface gum, an arabinogalactan comprising some 15% galacturonic acid residues that enhance viscosity [45,54–59]. It has been suggested that the polysaccharide gums found in lesquerella could be employed as an adhesive [22,23]. Apparently, these results confirm these earlier observations. The combination of the proteins (30–36%) with polysaccharide gums (15–20%) results in superior adhesive properties exhibited by LPC when compared to DDGS or PPC. Extraction of LPC by ethanol and hexane extraction resulted in composites (LPC/E-PW and LPC/H-PW) that had the highest flexural properties obtained in this study (Figure 1). The MOR and MOE of LPC/E and LPC/H were +31% and +29% and +31 and +25% greater than the MOR and MOE values of LPC-PW composite. Interestingly, the supercritical fluid extraction method resulted in a composite (LPC/CO_2-PW) that had inferior flexural properties to all other LPC composites. Further research would be necessary to clarify why this particular method failed to improve the adhesive properties of LPC.

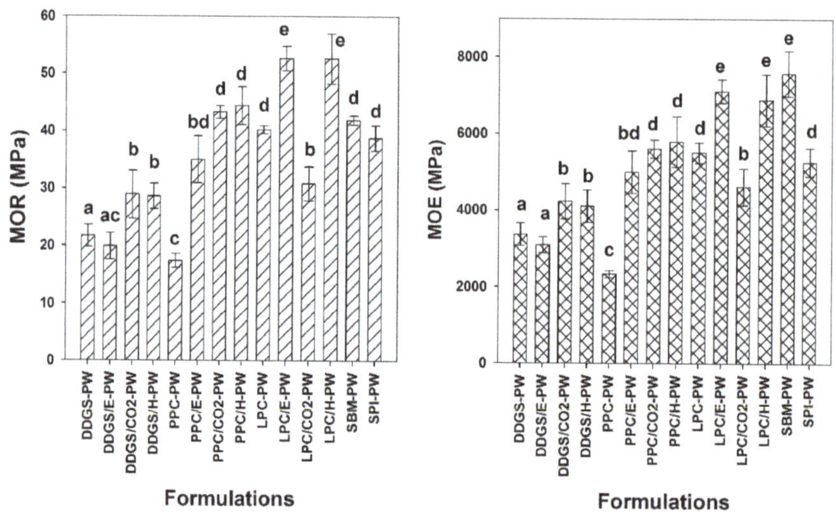

Figure 3. Mechanical properties of various panel formulations Means and standard errors are presented; treatment having different letters were significant ($P \leq 0.05$).

The highest MOE and MOR values were obtained from the LPC compared to the DDGS and PPC composites. For example, MOR and MOE values of LCP-PW were +85% and +163% greater than MOR and MOE values of DDGS-PW. This suggests that LCP may have some additional adhesive properties over those of the other two materials tested that not attributed to the protein content. This observed characteristic of the LPC matrix can be rationalized from earlier reported composition of the seed polysaccharide, namely, the seed surface gum, an arabinogalactan comprising some 15% galacturonic acid residues that enhance viscosity [45,54–59]. It has been suggested that the polysaccharide gums found in lesquerella could be employed as an adhesive [22,23]. Apparently, these results confirm these earlier observations. The combination of the proteins (30–36%) with polysaccharide gums (15–20%) results in superior adhesive properties exhibited by LPC when compared to DDGS or PPC. Extraction of LPC by ethanol and hexane extraction resulted in composites (LPC/E-PW and LPC/H-PW) that had the highest flexural properties obtained in this study (Figure 1). The MOR and MOE of LPC/E and LPC/H were +31% and +29% and +31 and +25% greater than the MOR and MOE values of LPC-PW composite. Interestingly, the supercritical fluid extraction method resulted in a composite (LPC/CO_2-PW) that had inferior flexural properties to all other LPC composites. Further research would be necessary to clarify why this particular method failed to improve the adhesive properties of LPC.

The extraction method employed to remove oils was found to profoundly affect the adhesive properties of DDGS and press cakes in the composites (Figure 3). For DDGS, supercritical fluid extraction and hexane extraction produced superior composites, compared composites employing matrices obtained from no extraction, or ethanol extraction. In the case of Pennycress, supercritical fluid extraction and hexane again gave rise to superior composites compared to unextracted matrix or ethanol extraction. However, PPC/E-PW composites obtained from ethanol extraction were clearly superior to untreated controls (PPC-PW). Superior mechanical properties were obtained in Lesquerella composites that were derived from press cakes subjected to ethanol and hexane extraction compared to untreated control composites. Extraction of LPC by supercritical fluid extraction resulted in an inferior composite (LPC/CO_2-PW) compared to composites derived from the other treatments. These results suggest hexane extraction overall appears to be the most reliable method to treat DDGS and press cakes, while CO_2 and ethanol extraction may result in an inferior composite depending on the starting material employed.

A comparison between MOR and MOE values of composites produced from a matrix of DDGS, PPC, or LPC and composites of a matrix using SBM or SPI is shown in Figure 3. The flexural properties of the SBM-PW composite were similar to several composites (e.g., PPC/CO_2-PW, PPC/H-PW, and LPC-PW). SBM and SPI matrix composites were superior to all DDGS composites (Figure 3). For example, MOR and MOE MPa values of SBM-PW were 41.8 ± 0.8 and 7.575 ± 578, respectively, compared to MOR and MOE MPa values of DDGS/H-PW, which exhibited 28.6 ± 2.2 and 4.228 ± 427, respectively. It should be noted that SBM contains ~50% protein, while the hexane extracted DDGS, PPC, and LPC contained 29%, 31%, and 36% protein, respectively (Table 2). This suggests that protein content may not be the only factor involved in the adhesive properties of these matrix materials. The SPI-PW composite contained ~45% proteins but did not exhibit greater flexural properties than the SBM-PW composite, which contained 24% protein or several other press cake composites (Figure 3). Several PPC and LPC composites (i.e., PPC/CO_2-PW and LPC/E-PW) were found to exceed the SPI-PW composite in terms of flexural properties. Several investigators have demonstrated the adhesive properties of various commercially available soy flours and products [30,32,48]. It is the contention of the authors that alternative agriculture by products—namely, DDGS, PPC, and LPC—also have adhesive properties that may compete with soy flours.

It should be noted that those composite panels that exhibited higher densities and were thinner also exhibited higher mechanical properties compared to composite panels that had lower densities and were thicker (Figure 3, Table 6). All composite materials contained the same amount of materials and were fabricated in the same manner. The matrix material and method of extraction were responsible for the physical properties of the composite panel, which in turn influenced its mechanical properties. The panels employed in this study were unique in that they contained a high proportion of matrix material versus wood, i.e., 50:50 mixture. The matrix itself also contains a considerable amount of lignocellulosic material (~43–52%); thus, the the composite is composed of 73–79% lignocellulose and 13.3–17.5% protein (Table 2). Our lignocellulosic panels had a density between 986 to 1.256 kg/m^3. Commercial panels usually contain 85–90% wood, with the remaining components being adhesives and additives. High density fiber boards (HDF) panels have a density (900–1000 kg/m^3), and medium density fiberboards (MDF) have a density of (600–1000 kg/m^3). HDF and MDF are generally marketed in thickness varying from 3 to 12 mm in thickness. The panels employed in this study roughly correspond to HDF boards.

3.3. Dimensional Stability

It is important to ascertain the dimensional stability of our engineered panels and compare them to panels employing recognized adhesives [60,61]. The water absorption (WA) and thickness swelling (TS) of the various composites are shown in Table 6. Following 24 h of soaking composites, TS increased from 29 to 83%, while WA increased from 28 to 84% depending on the composite composition. Generally, there was a close association between the thickness and density of the TS and

WA values (Tables 6 and 7). However, the extraction method and matrix type also were contributing factors (Table 6). For example, comparing some DDGS composites, DDGS/H-PW had a thickness of 3.55 ± 0.10 mm, and a density of 1.188 ± 47 kg/m^3 exhibited TS of 42 ± 4% and WA of 37 ± 3%, while DDGS/E-PW had a thickness of 4.28 ± 0.28 mm, and a density of 986 ± 280 kg/m^3 exhibited 83 ± 4% TS and 84 ± 12% WA. Interestingly, the soy flour composites, SPI-PW and SBM-PW, exhibited relatively high TS and WA comparable to other composites tested. These results suggest that high protein content in these composites imbibes water, which adversely affects the dimensional stability. The employment of hexane extraction was found to be the most effective treatment for the various composites to improve the mechanical, physical, and dimensional stability properties (Figure 3, Table 6). Soy flour adhesives are noted to be effective in terms of their strength when compared to synthetic adhesives but exhibit poor durability, since they are susceptible to water damage [30,32,48,49].

Table 6. Dimensional Stability properties of composites soaked in distilled water for 24 h. Means and standard errors are presented; treatment having different letters in the same columns were significant ($p \leq 0.05$).

Formulations	Thickness (mm)	Density (kg/m^3)	TS (%)	WA (%)
DDGS-PW	3.84 ± 0.09 a	1.133 ± 65 a	57 ± 8 a	48 ± 5 a
DDGS/E-PW	4.24 ± 0.28 ab	986 ± 280 a	83 ± 4 b	84 ± 12 b
DDGS/CO_2-PW	3.67 ± 0.10 ac	1.156 ± 39 a	60 ± 4 a	54 ± 4 a
DDGS/H-PW	3.55 ± 0.07 c	1.188 ± 47 a	42 ± 4 c	37 ± 3 c
PPC-PW	3.44 ± 0.08 c	1.256 ± 19 a	61 ± 4 a	51 ± 3 a
PPC/E-PW	3.67 ± 0.18 ac	1.182 ± 43 a	61 ± 9 a	59 ± 11 a
PPC/CO_2-PW	3.39 ± 0.05 c	1.290 ± 18 b	31 ± 3 d	30 ± 3 d
PPC/H-PW	3.28 ± 0.03 d	1.312 ± 28 b	29 ± 4 d	28 ± 3 d
LPC-PW	3.38 ± 0.03 cd	1.296 ± 14 b	55 ± 4 a	45 ± 7 a
LPC/E-PW	3.50 ± 0.05 ac	1.239 ± 28 ab	56 ± 7 a	48 ± 6 a
LPC/CO_2-PW	4.11 ± 0.07 b	1.088 ± 117 a	62 ± 4 a	81 ± 15 b
LPC/H-PW	3.35 ± 0.04 cd	1.266 ± 19 ab	44 ± 6 ac	40 ± 3 c
SBM-PW	3.29 ± 0.04 d	1.262 ± 12 a	51 ± 2 a	49 ± 5 e
SPI-PW	3.70 ± 0.01 a	1.201 ± 18 a	73 ± 4 a	68 ± 5 e

Pearson correlation coefficients comparing the physical properties (thickness and density) of composites with their response to water (TS and WA) are presented in Table 7. There were high correlations between composite panel thickness and TS (0.775) and thickness and WA (0.898). The thicker the panel, the greater the TS and WA values. High negative correlations occur between panel density and TS (−0.753) and density and WA (−0.849). These correlations indicate a close relationship between the thickness and density of the panel and its dimensional stability.

Table 7. Pearson correlation coefficient values for the physical dimensions and dimensional stability [1].

Correlations	Thickness (mm)	Density (kg·m^{-3})	TS (%)	WA (%)
Thickness (mm)	-	−0.959	0.775	0.898
Density (kg·m^{-3})	−0.959	-	−0.753	−0.849
TS (%)	0.775	−0.753	-	0.892
WA (%)	0.898	−0.849	0.892	-

[1] All values were significant at $p = 0.05$ employing 5 observations.

Frihart et al. [48] noted that for a new adhesive to be accepted commercially, it must satisfy several requirements: (1) it must be abundant and have consistent properties; (2) it should be used with the typical industrial processing conditions and equipment; (3) it must have a similar viscosity compatible with existing equipment for specific processes; (4) it must not alter with the moisture content of the wood product during processing; and (5) it must provide a finished product of adequate strength and stiffness as defined by industry standards. In the wood industry, three main classes of fiberboard material classification are: interior non-structural, exterior non-structural, and structural [62–64]. In our case, the lignocellulosic panels produced from various agricultural by-products satisfy some of these commercial requirements, especially in terms of their flexural properties, and as such are comparable to soybean flours. However, the dimensional stability of the composite panels presented in this study precludes their use in exterior applications. According to the European standards for the nominal properties of PB, MDF, and HDF, TS values are 35% and less (typically 10–25%). Only two composites (PPC/CO_2-PW and PPC/H-PW) were able to achieve TS values that were of these dimensional stability standards. European standards for HDF nominal MOR and MOE properties vary from 30–44 MPa to 2500–4500 MPa, respectively [62]. Several of the LC panels fabricated in this study satisfy these property requirements (Figure 3). In addition, these agricultural by-product adhesives are abundant (or could be if commercialized), can be employed as a powder (as soy flour often is), and confer strength/stiffness in composite panels. The dimensional stability results suggest that these agricultural adhesives should be restricted to interior non-structural uses. More research is necessary to enhance the use of these agricultural by-products to make them commercially acceptable. However, these results are a promising beginning.

4. Conclusions

Seed meal by-products generated from dry milling (DDGS) and mechanical pressing of oil seeds (PPC and LPC) have been found to have adhesive properties. Using DDGS, PPC, and LPC as an adhesive/resin with wood lignocellulosic, composites panels were fabricated that exhibited high mechanical properties. Further, the extraction of residual oils enhanced the adhesive properties of these by-products. The type of extraction method to remove oils from DDGS, PPC, or LPC profoundly affected the mechanical, physical, and dimensional stability of subsequent composites generated. Overall, the Soxhlet extraction method using hexane was most effective at removing oil, giving rise to a matrix material that produced composites with higher mechanical properties compared to the other extraction methods. Composites containing the LPC matrix had greater mechanical properties than DDGS and PPC. Several DDGS, PPC, or LPC composites were found to have more similar mechanical properties than they were when using soy flour matrix materials. There was a close relationship between composite panel thickness and density and mechanical properties. Composite panels that were thinner with higher densities usually exhibited higher mechanical properties. Dimensional stability (i.e., thickness swelling and water absorbance) was found to be correlated with panel thickness and density. Overall, composite panels from hexane-treated matrix materials exhibited better dimensional stability than composite panels that were not. Further, several composite panels employing DDGS, PPC, or LPC exhibited better dimensional stability than composite panels employing soy flours.

Author Contributions: Brent Tisserat, Fred Eller, and Rogers Harry-O'kuru conceived and designed the experiments; Rogers Harry-O'kuru and Fred Eller conducted extraction experiments; Brent Tisserat prepared materials and panels and conducted mechanical testing; Brent Tisserat and Fred Eller analyzed the data; Brent Tisserat and Rogers Harry-O'kuru wrote the paper.

Acknowledgments: The authors are grateful to Jeffrey Teel and Nicholas Montesdeoca, ARS-USDA for technical assistance and Roque L. Evangelista for pennycress and lesquerella press cakes. Mention of trade names or commercial products in this publication is solely for the purpose of providing specific information and does not imply recommendation or endorsement by the US Department of Agriculture. USDA is an equal opportunity provider and employer. Funding was provided by USDA-ARS.

Conflicts of Interest: The authors declare no conflict of interest.

References

1. Irwin, S.; Good, D. Ethanol production profits: The risk from lower prices of distillers grains. *Farmdoc Daily* **2015**, *5*, 46. Available online: http://farmdocdaily.illinois.edu/2015/03/ethanol-production-profits-risk-of-distillers-grains.html (accessed on 1 December 2017).
2. Roetheli, J.C.; Carlson, K.D.; Kleiman, R.; Thompson, A.E.; Dierig, D.A.; Glaser, L.K.; Blase, M.G.; Goddell, J. *Lesquerella as a Source of Hydroxyl Fatty Acids for Industrial Products*; Growing Industrial Materials Series; USDA-CSRS Office of Agricultural Materials: Washington, DC, USA, 1991.

3. Koeleman, E.; RELX Group. DDGS: Be Aware of Mycotoxins. All about Feed. 2016. Available online: http://www.allaboutfeed.net/Mycotoxins/Articles/2016/6/DDGS-be-aware-of-mycotoxins-2822263W/ (accessed on 5 January 2018).
4. Tilstra, H. *Corn DDGS: A Feed Industry Perspective*; Land O'Lakes Farmland Feed, LLC: Luverne, MN, USA, 2012. Available online: http://www.ddgs.umn.edu/sites/biodieselfeeds.cfans.umn.edu/files/cfans_asset_414268.pdf (accessed on 5 January 2018).
5. Fabiosa, J.F. *Not All DDGS Are Created Equal: Nutrient-Profile-Based Pricing to Incentivize Quality*; Iowa State University: Ames, IA, USA, 2008. Available online: http://ageconsearch.umn.edu/bitstream/44755/2/08-WP_481.Fabiosa.pdf (accessed on 3 December 2017).
6. Bamualim, A.; Nurhayati, I.S.; Tangendjaja, B. *DDGS Feeding Trial on Dairy Cattle in INDONESIA*; U.S. Grains Council: Washington, DC, USA, 2009. Available online: http://www.ddgs.umn.edu/sites/biodieselfeeds.cfans.umn.edu/files/cfans_asset_414584.pdf (accessed on 5 January 2018).
7. Wisner, R.; Estimated U.S. Dried Distillers Grains with Solubles (DDGS) Production & Use. 2016. Available online: http://www.extension.iastate.edu/agdm/crops/outlook/dgsbalancesheet.pdf (accessed on 5 January 2018).
8. USDA Market News Service. *National Weekly Feedstuff Wholesale Prices*; USDA Market News Service: Minneapolis, MN, USA, 2016. Available online: www.ams.usda.gov/mnreports/ms_gr852.txt (accessed on 5 January 2018).
9. Bowen, N. China Bans U.S. Distillers Grains Imports. National Association of State Departments of Agriculture: Arlington, VA, USA, 2014. Available online: http://www.nasda.org/News/27570.aspx (accessed on 5 January 2018).
10. Bowers, E.L.; Munkvold, G.P. Fumonisins in conventional and transgenic, insect-resistant maize intended for fuel ethanol production: Implications for fermentation efficiency and DDGS Co-product quality. *Toxins* **2014**, *6*, 2804–2825. [CrossRef] [PubMed]
11. Dillivan, K.U.S. Corn and DDGS Exports. University News Release. Available online: http://www.agweb.com/article/us-corn-and-ddgs-exports-university-news-release/?print=y (accessed on 4 January 2018).
12. Donley, A. Keeping Mycotoxins Out of DDGS. 2010. Available online: http://www.world-grain.com/News/Archive/Keeping-mycotoxins-out-of-DDGS.aspx?p=1 (accessed on 4 January 2018).
13. Irwin, S.; Good, D. Understanding the pricing of Distillers' grains Solubles. *Farmdoc Daily* **2013**, *3*, 133. Available online: http://farmdocdaily.illinois.edu/2013/07/understanding-pricing-distillers-grain.html (accessed on 3 December 2017).
14. Salywon, A.M.; Dierig, D.A.; Rebman, J.P.; De Rodríquez, D.J. Evaluation of new *Lesquerella* and *Physaria* (Brassicaceae) oilseed germplasm. *Am. J. Bot.* **2005**, *92*, 53–62. [CrossRef] [PubMed]
15. Wang, G.; McCloskey, W.; Foster, M.; Dierig, D. *Lesquerella: A Winter Oilseed Crop for the Southwest*; Report AZ1520, Arizona Cooperative Extension; University AZ: Tucson, AZ, USA, 2010. Available online: https://extension.arizona.edu/pubs/az1520.pdf (accessed on 2 February 2018).
16. Dierig, D.A.; Wang, G.; McCloskey, W.B.; Thorp, K.R.; Isbell, T.A.; Ray, D.T.; Foster, M.A. Lesquerella: New crop development and commercialization in the U.S. *Ind. Crops Prod.* **2011**, *34*, 1381–1385. [CrossRef]
17. Hojilla-Evangelista, M.P.; Selling, G.W.; Berhow, M.A.; Evangelista, R.L. Preparation, composition and functional properties of pennycress (*Thlaspi arvense* L.) seed protein isolates. *J. Am. Oil Chem. Soc.* **2015**, *92*, 905–914. [CrossRef]
18. Smith, M. Pennycress. AgMRC, Iowa State University: Ames, IA, USA, 2015. Available online: http://www.agmrc.org/commodities-products/grains-oilseeds/pennycress/ (accessed on 4 December 2017).
19. Vaughn, S.F.; Deppe, N.A.; Berhow, M.A.; Evangelista, R.L. Lesquerella press cake as an organic fertilizer for greenhouse tomatoes. *Ind. Crops Prod.* **2010**, *32*, 164–168. [CrossRef]
20. Vaughn, S.F.; Isbell, T.A.; Weisleder, D.; Berhow, M.A. Biofumigant compounds released by field pennycress (*Thiaspi arvense* L.) seedmeal. *J. Chem. Ecol.* **2005**, *31*, 167–177. [CrossRef] [PubMed]
21. Reifschneider, L.; Tisserat, B.; Harry O'kuru, R. Mechanical properties of high density polyethylene—Pennycress press cake composites. In Proceedings of the 71st Annual Technical Conference of the Society of Plastics Engineers, Cincinnati, OH, USA, 22–24 April 2013; Volume 1, pp. 679–683.
22. Harry O'kuru, R.E.; Carriere, C.J.; Wing, R.E. Rheology of modified *Lesquerella* gum. *Ind. Crops Prod.* **1999**, *10*, 11–20. [CrossRef]

23. Vaughn, S.F.; Kenar, J.A.; Felker, F.C.; Berhow, M.A.; Cermak, S.C.; Evangelista, R.L.; Fanta, G.F.; Behle, R.W.; Lee, E. Evaluation of alternatives to guar gum as tacksifiers for hydromulch and as clumping agents for biodegradable cat litter. *Ind. Crops Prod.* **2013**, *43*, 798–801. [CrossRef]
24. Harry O'kuru, R.; Gordon, S.H.; Klokkenga, M. Bio-generation of succinic acid by fermentation of *Physaria fendleri* seed polysaccharides. *Ind. Crops Prod.* **2015**, *77*, 116–122. [CrossRef]
25. Pahkasalo, T.; Aurenhammer, P.; Gaston, C. Value-Added Wood Products Markets, 2011–2012. In Geneva Timber and Forest Study Paper 30, Forest Products Annual Market Review, 2011–2012–UN/FAO, Geneva–New York. 2012; pp. 129–140. Available online: http://www.unece.org/fileadmin/DAM/timber/publications/12.pdf (accessed on 5 January 2018).
26. Ansell, M. *Wood Composites*, 1st ed.; Woodhead Publishing Series; Elsevier Ltd.: London, UK, 2015; ISBN 9781782424523.
27. Amaral-Labat, G.A.; Pizzi, A.; Goncalves, A.R.; Celzard, A.; Rigolet, S.; Rocha, G.J.M. Environmental-friendly soy four-based resins without formaldehyde. *J. Appl. Polym. Sci.* **2008**, *108*, 624–632. [CrossRef]
28. Gu, K.; Huang, J.; Li, K. Preparation and evaluation of particle board bonded with a soy flour-based adhesive with a new curing agent. *J. Adhes. Sci. Technol.* **2013**, *27*, 2053–2064. [CrossRef]
29. Formaldehyde. National Institude Enivronmental Health Science. 2018. Available online: https://www.niehs.nih.gov/health/topics/agents/formaldehyde/index.cfm (accessed on 2 February 2018).
30. Frihart, C.R.; Hunt, C.; Birkeland, M.J. Chapter 16: Soy proteins as wood adhesives. In *Recent Advances in Adhesion Science and Technology*; Gutowski, W., Dodiuk, H., Eds.; CRC Press: Boca Raton, FL, USA, 2014; pp. 277–294. [CrossRef]
31. Liu, Y.; Li, K. Development and characterization of adhesives from soy protein for bonding wood. *Int. J. Adhes. Adhes.* **2007**, *27*, 59–67. [CrossRef]
32. Mo, X.; Hu, J.; Sun, X.S.; Ratto, J.A. Compression and tensile strength of low-density straw-protein particleboard. *Ind. Crops Prod.* **2001**, *14*, 1–9. [CrossRef]
33. Tisserat, B. Development of Composites Utilizing Resin/adhesives Obtained from Dried Distillers Grains and Solubles. U.S. Patent Application No. 15,651,149, 17 July 2017.
34. Tisserat, B.; Reifschneider, L.; Harry O'Kuru, R.; Finkenstadt, V.L. Mechanical and thermal properties of high density polyethylene distillers grains with solubles composites. *Biores* **2013**, *8*, 59–75. [CrossRef]
35. Baumler, E.R.; Carrin, M.E.; Carelli, A.A. Extraction of sunflower oil using ethanol as solvent. *J. Food Eng.* **2016**, *178*, 190–197. [CrossRef]
36. Kwiatkowski, J.R.; Cheryan, M. Extraction of oil from ground corn using ethanol. *J. Am. Oil Chem. Soc.* **2002**, *79*, 825–830. [CrossRef]
37. Herrero, M.; Mendiola, J.A.; Cifuentes, A.; Ibáñez, E. Supercritical fluid extraction: Recent advances and applications. *J. Chrom. A* **2017**, *16*, 2495–2511. [CrossRef] [PubMed]
38. Hojilla-Evangelista, M.P.; Evangelista, R.L.; Isbell, T.A.; Selling, G.W. Effects of cold-pressing and seed cooking on functional properties of protein in pennycress (*Thlaspi arvense* L.) seed and press cakes. *Ind. Crops Prod.* **2013**, *45*, 223–229. [CrossRef]
39. Friedrich, J.P.; List, G.R. Characterization of soybean oil extracted by supercritical carbon dioxide and hexane. *J. Agric. Food Chem.* **1982**, *30*, 192–193. [CrossRef]
40. European Committee for Standardization. *Wood-Based Panels—Determination of Modulus of Elasticity in Bending and of Bending Strength*; EN 310:1993; European Committee for Standardization: Brussels, BE, USA, 1993; ISBN 0580210561.
41. European Committee for Standardization. *Particleboards and Fibreboards—Determination of Swelling in Thickness after Immersion in Water*; EN 317:1993; European Committee for Standardization: Brussels, BE, USA, 1993; ISBN 0580211363.
42. European Committee for Standardization. *Wood-Based Panels—Determination of Density*; EN 323:1993; European Committee for Standardization: Brussels, BE, USA, 1993; ISBN 0580210588.
43. Wall, J.S.; Paulis, J.W. Corn and sorghum grain proteins. In *Advances in Cereal Science and Technology*; Pomeranz, Y., Ed.; AACC: St. Paul, MN, USA, 1985; Volume II, pp. 142–219, ISBN 13 978-0913250082.
44. Selling, G.W.; Hojilla-Evangelista, M.P.; Evangelista, R.L.; Isbell, T.; Price, N.; Doll, K.M. Extraction of proteins from pennycress seeds and press cake. *Ind. Crops Prod.* **2013**, *41*, 111–119. [CrossRef]
45. Wu, Y.V.; Hojilla-Evangelista, M.P. *Lesquerella fendleri* protein fractionation and characterization. *J. Am. Oil Chem. Soc.* **2005**, *82*, 53–56. [CrossRef]

46. Karr-Lilienthal, L.K.; Bauer, L.L.; Utterback, P.L.; Zinn, K.E.; Frazier, R.L.; Parsons, C.M.; Fahey, G.C. Chemical composition and nutritional quality of soybean meals prepared by extruder/expeller processing for use in poultry diets. *J. Agric. Food Chem.* **2006**, *54*, 8108–8114. [CrossRef] [PubMed]
47. Wall, J.S.; Huebner, F.R. Adhesion and Cohesion. In *Protein Functionality in Foods*; Cherry, J.P., Ed.; ACS: Washington, DC, USA, 1981; pp. 111–130. Available online: https://naldc.nal.usda.gov/download/26684/PDF (accessed on 5 January 2018).
48. Frihart, C.R.; Birkeland, M.J. Chapter 8: Soy properties and soy wood adhesives. In *Soy-Based Chemicals and Materials*; ACS: Washington, DC, USA, 2014; pp. 167–192. [CrossRef]
49. Li, N.; Wang, Y.; Tilley, M.; Bean, S.R.; Wu, X.; Sun, X.S.; Wang, D. Adhesive performance of sorghum protein extracted from sorghum DDGS and flour. *J. Polym. Environ.* **2011**, *19*, 755–765. [CrossRef]
50. Li, N.; Qi, G.; Sun, X.S.; Xuk, F.; Wang, D. Adhesion properties of Camelina protein fractions isolated with different methods. *Ind. Crops Prod.* **2015**, *69*, 264–272. [CrossRef]
51. Nordqvist, P.; Nordgren, N.; Khabbaz, F.; Malmström, E. Plant proteins as wood adhesives: Bonding performance at the macro- and nanoscale. *Ind. Crops Prod.* **2013**, *44*, 246–252. [CrossRef]
52. Nordqvist, P.; Khabbaz, F.; Malmström, E. Comparing bond strength and water resistance of alkali-modifed soy protein isolate and wheat gluten adhesives. *Int. J. Adhes. Adhes.* **2010**, *30*, 72–79. [CrossRef]
53. Winkler, J.K.; Rennick, K.A.; Eller, F.J.; Vaughn, S.F. Phytosterol and tocopherol components in extracts of corn distiller's dried grain. *J. Agric. Food Chem.* **2007**, *55*, 6482–6486. [CrossRef] [PubMed]
54. Tookey, H.L.; Lohmar, R.L.; Wolff, I.A.; Jones, Q. New sources of seed mucilages. *J. Agric. Food Chem.* **1962**, *10*, 131–133. [CrossRef]
55. Tookey, H.L.; Jones, Q. New sources of water-soluble seed gums. *Econ. Bot.* **1965**, *19*, 165–174. [CrossRef]
56. Mikolajczak, K.L.; Earle, F.R.; Wolff, I.A. Search for new industrial oils. VI. Seed oils of the genus Lesquerella. *J. Am. Oil Chem. Soc.* **1962**, *39*, 78–80. [CrossRef]
57. Holser, R.A.; Carriere, C.J.; Abbot, T.P. Rheological properties of lesquerella gum fractions recovered by aqueous extraction. *Ind. Crops Prod.* **2000**, *12*, 63–69. [CrossRef]
58. Bolker, H.I. *Natural and Synthetic Polymers*; Dekker: New York, NY, USA, 1974; ISBN 9780824710606.
59. Abbott, T.P.; Wu, V.Y.; Carlson, K.D.; Slodki, M.E.; Kleiman, R. Isolation and preliminary characterization of *Lesquerella fendleri* gums from seed, presscake and defatted meal. *J. Agric. Food Chem.* **1994**, *42*, 1678–1685. [CrossRef]
60. Zhong, A.; Sun, S.X.; Fang, X.; Ratto, J.A. Adhesion properties of soy protein with fiber cardboard. *J. Am. Oil Chem. Soc.* **2001**, *78*, 37–41. [CrossRef]
61. Garcia, R.A.; Cloutier, A.; Riedl, B. Dimensional stability of MD panels produced from heat-treated fibers. *Holzforschung* **2006**, *60*, 278–284. [CrossRef]
62. European Committee for Standardization. *Fibreboards—Specifications—Part 2: Requirements for Hardboards*; EN 622-2:1997; European Committee for Standardization: Brussels, Belgium, 1997; ISBN 0580494977.
63. European Committee for Standardization. *Particleboard—Specifications*; EN 312:2003; European Committee for Standardization: Brussels, Belgium, 2003.
64. European Committee for Standardization. *Fibreboards—Specifications—Part 5: Requirements for Dry Process Boards (MDF)*; EN 622-5:2006; European Committee for Standardization: Brussels, Belgium, 2006; ISBN 058049165X.

fibers

MDPI

Article

Physico-Mechanical, Thermal and Biodegradation Performance of Random Flax/Polylactic Acid and Unidirectional Flax/Polylactic Acid Biocomposites

Mahmudul Akonda [1], S. Alimuzzaman [2], D. U. Shah [3] and A.N.M. Masudur Rahman [2,*]

1 Department of Textile Engineering, Southeast University, Dhaka 1208, Bangladesh; mhakonda@yahoo.com
2 Department of Fabric Engineering, Bangladesh University of Textiles, Dhaka 1208, Bangladesh; sazaman_2006@yahoo.com
3 Centre for Natural Material Innovation, Department of Architecture, University of Cambridge, Cambridge CB2 1PX, UK; dus20@cam.ac.uk
* Correspondence: masudfabric@yahoo.com; Tel.: +880-15-5334-2607

Received: 11 August 2018; Accepted: 30 November 2018; Published: 10 December 2018

check for updates

Abstract: Fully biodegradable flax/polylactic acid (PLA) thermoplastic composites were fabricated by using random (nonwoven mat) and aligned (unidirectional yarn) flax fiber as reinforcements (39% flax by volume) and Polylactic acid (PLA) as matrix. Results revealed that the aligned flax fibers have a greater reinforcing effect due to the uniform distribution of load axially along the fiber length in the composite. The aligned flax/PLA and random flax/PLA showed the tensile strength of (83.0 ± 5.0) and (151.0 ± 7.0) MPa respectively and flexural strength of (130.0 ± 5.0) and (215.0 ± 7.2) MPa respectively. Young's modulus of (9.3 ± 1.5) and (18.5 ± 2.0) GPa and flexural modulus of (9.9 ± 1.0) and (18.8 ± 1.0) GPa was attained for the random and unidirectional fiber composites, respectively. It was also found that both composite constituents, fiber and matrix, were degradable if buried in compost soil (ready soil after composting process), which is a distinctive advantage of the new composite structures. Remarkably, the biodegradation property of aligned flax fiber composites was significantly lower than random mat composites, possibly due to the less water swelling behavior of the aligned fiber composites. After 120 days burial test, the aligned flax/PLA composite displayed the reduction of 19% mass, residual flexural strength and modulus decreased by 57 and 50% respectively, while the random mat composites exhibited the loss of 27% mass, residual flexural strength and modulus declined by 80% at the same period.

Keywords: PLA; flax; thermoplastic composites; mechanical properties; biodegradability; durability

1. Introduction

Composite materials based on natural and renewable resources have received amazing interest in recent years due to a drive towards sustainable manufacturing and materials end-of-life management. It is well known that typical high-performance composite components, such as carbon and glass fibers and petrochemical-based matrices, require high energy for production, release considerable CO_2 into the atmosphere, are not biodegradable and not considered to be environmentally-friendly [1–3]. Also, the extreme consumption of petroleum-based plastics, which prompt massive volumes of non-decomposable wastes origins a thoughtful exhaustion of landfill dimensions. The wakefulness of climbed waste complications on the environment has stimulated a new concentration in the area of materials science. Because of the growing environmental awareness in the world, it is a critical issue for scientists to study diverse substitutions to switch non-biodegradable materials, particularly for petroleum-based products [4–7]. Consequently, various types of fully biodegradable resources are being developed recently, as replacements for non-renewable plastics. For a sustainable future, environmentally friendly composite components (EFCs) with constituents that will be biodegradable at the end of usable life are of interest [8–10].

Biocomposites are the combination of biofibers with polymer matrices from both of non-renewable and renewable resources. Different types of biofibers such as jute [11], sisal [12], hemp [13], kenaf [14] and flax [15] have been investigated for reinforcing both thermoplastic and thermoset matrices. Biofibers are one of the major components of biocomposites and the fibrous material derived from plant sources. Biofiber composites are emerging as a sustainable alternate to glass fiber composites, mainly in packaging, automotive, building and consumer product industries and becoming one of the fastest rising additives for EFCs [16]. Among the commodity of biofibers, flax fibers show some unique mechanical properties. Other advantages such as production cost with low investment cost make it an interesting product for low-wage countries. Thermal recycling is also possible where glass causes problems in combustion furnaces. Flax fiber possesses low specific weight and higher stiffness than glass (Table 1). It also provides better thermal and acoustic insulation properties, especially as an automotive interior or construction material part [17,18].

Table 1. Comparison of glass fiber with some natural fiber properties [19,20].

Fiber Type	Density (g/m^3)	Tensile Strength (MPa)	Tensile Modulus (GPa)	Elongation at Break (%)	Moisture Absorption (%)
Glass	2.55	2000–3500	70–73	2.5–3	-
Flax	1.4–1.54	800–2000	60–85	1.2–4	7
Hemp	1.48	368–900	70	1.6	8
Jute	1.44–1.46	400–775	10–30	1.5–1.8	12
Sisal	1.33–1.5	350–700	22–38	2–7	11
Coir	1.25	220	6	15–25	10
Cotton	1.5–1.6	200–800	5.5–12.6	6–12	8–25

Although natural fiber reinforced thermoplastic composites are becoming more popular, particularly in the automotive industry but the options for the use of biodegradable thermoplastics are limited [21,22]. To obtain 'fully' biodegradable EFCs, the petrochemical-based matrix can be replaced by a vegetable-derived polymer, such as PLA. Among numerous ecofriendly polymers, PLA attracts particular attention for the improvement of engineering polymers [23,24]. PLA belongs to synthetic aliphatic polyesters (without branching between chains and strong entanglement). It is generally resulting from α-hydroxy acids, which are manufactured by means of monomer feedstock. PLA is produced through the condensation polymerization of a basic building block lactic acid (LA), which is generally made by microbial fermentation of carbohydrates (mainly hexose) in corn (glucose), cheese whey (lactose) and sugar cane (sucrose). Since PLA is compostable and derived from renewable, sustainable sources, it is regarded as a favorable material to improve collective solid waste disposal problems. However, contemporary applications of natural fiber/PLA thermoplastic materials are limited due to the mechanical properties of the composite [3,20,22,25,26].

To maximize the stiffness and strength of laminated composites, particularly for load-bearing applications, unidirectional reinforcements are the most preferable structures [27]. Usually, long natural fibers provide better mechanical properties to composite structures compared to short fiber reinforcements. However, the mechanical properties of composites also depend on other architectural parameters like fiber orientation, distribution, volume fraction and manufacturing effects (defects). As an example, flax mat/PP film sandwich structure shows poor mechanical properties compared to flax/PP commingled yarn composites due to poor fiber wettability. Similarly, due to damage of the fibers caused by needle punching during mat production process, needle-punched nonwoven mat of flax/PLA film structure provides reduced mechanical properties compared to flax yarn/PLA film structure [28]. Such parameters can also influence the water absorption and biodegradation properties of the composites.

The present research was focused on developing two different types of preforms such as random and aligned preforms produced from long flax fibers to manufacture flax/PLA composite. A novel air-laid technology was used to obtain fully-random flax reinforcements by avoiding needle-punching or film-stacking processes to maximize the reinforcing effect of fiber in nonwoven mat composites. In addition, a new commingled yarn structure from flax/PLA fiber blends was produced to fabricate composites. Thereafter, mechanical, water absorption and biodegradability performance of the biocomposite materials were evaluated.

2. Materials and Methods

2.1. Materials

Commercial grade flax (untreated) and short staple PLA fibers were sourced from Procotex and DS fibers, Belgium respectively. Some specifications of the used fibers and produced materials are presented in Table 2.

Table 2. Physical properties of flax, PLA fibers and their intermediate products.

Properties	Flax Fiber	PLA Fiber	Flax/PLA Nonwoven Mat	Flax/PLA Yarn
Fiber diameter (μm)	22.0 ± 1.0	28.0 ± 0.5	-	-
Linear density	6.6 ± 0.5 dtex	3.3 ± 0.5 dtex		250.0 ± 10.0 tex
Areal density	-	-	1000.0 ± 50.0 g/m^2	-
Mean fiber length (mm)	70.0 ± 17.0	70.0 ± 3.0	-	-
Tenacity (MPa)	750.0 ± 15.0	104.0 ± 5.0	-	-
Breaking elongation (%)	1.8 ± 0.5	2.9 ± 1.0	0.8 ± 1.0	3.5 ± 1.0
Melting temperature (°C)	-	171.0	-	-
Density (g/cm^3)	1.54	1.24	-	-

2.2. Evaluation of Fiber Property and Preform

Areal density (g/m^2) and the thickness of the preforms were tested. The areal density was determined according to ISO 9073-1: 1989 standard. The thickness was measured according to ISO 9073-2: 1997 standard by using Shirley thickness gauge under a pressure foot of 8 cm diameter and pressure of 250 gm.

To calculate the linear density of the fibers ISO 1973: 1996 standard was followed. According to the standard 10 fiber tufts having a mass of several milligrams were taken from the sample and the fibers of each tuft were brought into parallel arrangement. The middle part of each combed tuft was 50 mm in length. Five fibers were taken out from each of ten bundles in turn, so as to form a bundle of 50 fibers and 10 of these bundles were made. These bundles were weighted individually using the balance. The mean linear density of fiber in each bundle was calculated by Equation (1).

$$\rho_{l,b} = \frac{m_b}{n_f \times l_f} \times 10^4 \quad (1)$$

where, $\rho_{l,b}$ is the mean linear density of the fiber in each bundle in decitex (dtex); m_b is the mass of fiber bundle in mg; n_f is the number of fibers in the bundle; l_f is the length of the individual fibers in the bundle in mm.

The areal density was measured by the following equation:

$$M = \frac{m}{A} \times 10^6 \quad (2)$$

where, M is the areal density in g/m^2; m is the mass per sample area in gm; A is the area of the sample in mm^2.

2.3. Production of Flax/PLA Preforms

2.3.1. Aligned Flax Fiber Preforms

Flax fibers (sliver form) were commingled with staple PLA fibers at a ratio of 50:47 (wt.%) to produce a low twisted yarn structure. Conventional spinning process was used to produce flax/PLA yarn. The provided sliver form of flax and staple PLA fiber was passed through the carding and drawing process, rove is obtained by doubling of several number of slivers and finally yarn is produced by inserting 40 turn per meter (considered as low twist) in spinning frame. Additional 3% PLA were used as wrappers with a wrapping attachment in spinning frame to achieve 50% mass fraction of the PLA fibers in final yarns. The wrapper PLA was used as filament (150 dtex) form which was also collected from DS fibers, Belgium.

2.3.2. Randomly Aligned Flax Fiber Preforms

From the literature survey, it is found that most nonwoven preforms are made from stitch bonded mats. The stitch bonded nonwovens are based on needle punching, which can cause fiber damage [29]. The thermal bonding route used in this research work does not suffer from this drawback. The matrix material is also used as binding material here. The web is consolidated by using a hot pressing instrument.

Nonwoven webs (250 g/m^2) were prepared from flax/PLA blends using a mass ratio of 50:50 (wt.%) by means of an air laying process [30]. An air laying web forming machine generally consists of feeding, carding, stripping, transporting and condensing units which ensures the isotropic fiber orientation distribution, leading to isotropic composites. The mechanical properties of the composites are similar in all direction due to the isotropic fiber distribution [31]. Four layers of webs were piled-up together and the PLA fibers were partly melted to make a thicker 'batt' (mat) with an areal density of (1000 $\pm$ 50) g/m^2 to be used in the hot-press molding process. To ensure uniform blends, flax and PLA fibers were pre-opened manually prior to feeding into the nonwoven process. Nonwoven preform manufacturing technique is shown in Figure 1.

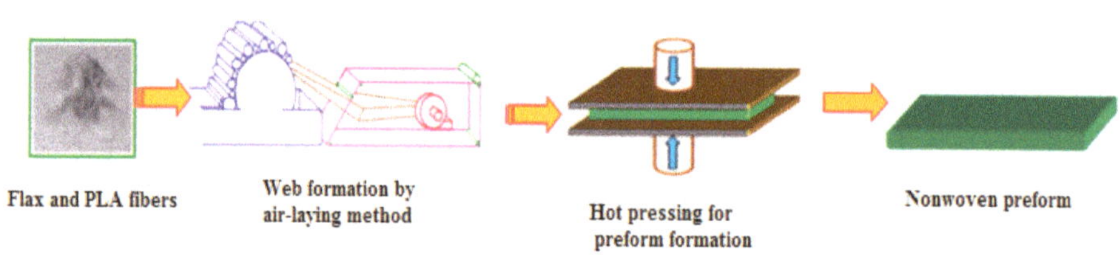

Figure 1. Schematic diagram of random flax preform fabrication process route.

2.3.3. Aligned Flax Fiber Composites

To obtain aligned fiber architecture in the composites, commingled yarns of flax/PLA were wound onto a steel frame (60 mm $\times$ 90 mm) unidirectionally. Six layers of yarn were subsequently compressed and consolidated using a hot compression molding process. The temperature was maintained 185 °C by keeping the pressure of 2 MPa for 5 min to produce 2.1 mm thick panels. The aligned (unidirectional) flax composites were termed as UD flax/PLA composites.

2.3.4. Randomly Aligned Flax Fiber Composites

To achieve random fiber architecture in the composites, two layers of nonwoven mats were used to produce 2.3 mm thick laminates. The molding parameters such as temperature and time were kept same as used for aligned laminates but higher pressure (4 MPa) was applied to compact the bulky mats and the produced composites were designated as NM flax/PLA composites.

2.4. Characterization of Composites

2.4.1. Mechanical Testing

The tensile properties of composite panels were measured in accordance with ISO 527-4:1997. The tests were conducted on a universal Instron testing machine, model 5569 with a 10 KN load cell at an extension rate of 2.0 mm/min. The Young's modulus was determined by considering a linear regression in the strain range of 0.0–0.2% [32]. Unidirectional composite samples were tested in the fiber direction. The flexural (3-point bending) tests were carried out according to ISO 14125:1998 by using another universal Instron testing machine, model 4411 with 5 KN load cell at a constant crosshead speed of 2.0 mm/min. For these tests the specimen size was maintained (60 mm × 25 mm × 2.1 mm) for UD flax/PLA and (60 mm × 25 mm × 2.3 mm) for NM flax/PLA composites. All tests were performed at 23 ± 2 °C temperature and RH 50 ± 5%. Five samples of each type were tested and the final values were taken as the average value of five measurements.

2.4.2. Density, Fiber Volume Fraction and Void Content Measurement

In order to calculate the constituents of the composite, it is necessary to separate the reinforcement and the matrix. For this purpose, the resin digestion method ISO 14127: 2008 was used. In this method, Dichloromethane (DCM) was used to digest the PLA matrix [33]. The desiccated mass of the specimen was determined before and after the matrix digestion. The composite specimens were soaked in a bath at room temperature to digest the matrix. Sintered glass crucibles, oven, desiccators, DCM, conical flax and weighing balance were used in this experiment. The density of each specimen was calculated from the mass and volume. For these experiments, the average values were calculated from five measurements and considered as final value.

The fiber content, void content and flax volume fraction were calculated using the following Equations (3)–(7):

$$W_f\ (\%) = \frac{m_3 - m_1}{m_2 - m_1} \times 100 \tag{3}$$

where, W_f is the fiber content as a percentage of the initial mass; m_1 is the sintered glass crucible (gm); m_2 is the initial mass of the specimen and glass crucible (gm); m_3 is the final total mass (gm) of the crucible and residue after digestion.

$$W_r = 100 - W_f \tag{4}$$

where, W_r is the PLA content as a percentage of the initial mass; W_f is the fiber content as a percentage of the initial mass.

$$V_f = W_f \times \frac{\rho_c}{\rho_f} \tag{5}$$

where, V_f is the fiber content as a percentage of the initial volume, W_f is the fiber content as a percentage of the initial mass, ρ_c is the density of the test specimen (g/m^3), ρ_f is the density of flax (g/m^3).

$$V_r = \left(100 - W_f\right) \times \frac{\rho_c}{\rho_r} \tag{6}$$

where, V_r is the PLA content as a percentage of the initial volume; W_f is the flax content as a percentage of the initial mass; ρ_c is the density of the test specimen (g/m^3); ρ_r is the density of PLA (g/m^3).

$$V_0 = 100 - \left[W_f \times \frac{\rho_c}{\rho_f} + \left(100 - W_f\right) \times \frac{\rho_c}{\rho_r}\right] \tag{7}$$

where, V_0 is the void content as a percentage of the initial volume, W_f is the fiber content as a percentage of the initial mass, ρ_c is the density of the test specimen (g/m^3), ρ_f is the density of flax fiber (g/m^3), ρ_r is the density of PLA (g/m^3).

2.5. Differential Scanning Calorimetry (DSC)

The DSC technique determines the quantity of heat either absorbed or released when a substance undergoes a physical or chemical change. Several parameters can be estimated by performing a DSC scan. These parameters include the glass transition temperature (T_g), melting temperature (T_m), melting enthalpy (ΔH_m) and crystalline level. The DSC measurements were carried out by TA Instrument DSC Q100. The heating and cooling rate were 10 °C/min. The samples were first scanned from −20 to 220 °C, followed by cooling to −20 °C and then reheating up to 220 °C in a second scan. T_m was determined from the maximum region of the endothermic melting peak and T_g was an inflection temperature from the baseline of the first heating cycle. The degree of crystallinity (X_c) was calculated by Equation 8 using data from the DSC curve of the first heating cycle.

$$X_c(\%) = 100 \times \frac{\Delta H_m}{\Delta H_f} \tag{8}$$

where ΔH_m is the heat of fusion of the neat PLA and its composites, ΔH_f is the heat of fusion for 100% crystalline PLA (93.7 J/g).

2.6. Water Absorption Test

Water absorption determines the water swelling behavior of the composites. Water absorption test of the composite samples was carried out according to ISO 62:2008 (method 1). The samples (60 mm × 60 mm × 2.1 mm for UD and 60 mm × 60 mm × 2.3 mm for NM panel) were immersed in a beaker containing 300 mL distilled water for 60 days at room temperature. Before immersion, the samples were dried in a vacuum oven at 50 ± 2 °C temperature for 24 h, cooled in a desiccator and then immediately weighted (m_1). During the test, samples were withdrawn from the water at 20 days intervals, gently blotted with tissue paper to remove excess water from the surface, immediately weighed (m_2) and again dipped into the water. Each m_1 and m_2 value was an average value of three measurements. The gained mass percentage C (%) of the specimens due to water absorption was calculated using Equation (9).

$$C\ (\%) = 100 \times \frac{m_2 - m_1}{m_1} \tag{9}$$

2.7. Biodegradability Test

Biodegradability is one of the most important properties of the biocomposites. Soil burial test is the most widely used method to evaluate the biodegradability of the composite material. Different types of soil and compost have been used for biodegradation test. Kim et al. [34] showed that the biodegradability of the composite materials in compost soil is superior to that in a natural soil environment. Burial test was carried out in a flower pot containing Miracle Gro. moisture controlled compost soil [34,35]. The samples (60 mm × 25 mm × 2.1 mm for UD flax and 60 mm × 25 mm × 2.3 mm for NM flax panel) were buried randomly in compost soil for 4 months at a depth of 12–15 cm from the soil surface to ensure aerobic degradation [36]. The pots were covered with plastic film to avoid water evaporation from the compost surface. The P^H was maintained at 7 using a moisture meter and moisture content was preserved 40–50% by sprinkling water at an interval of 2 days. Biodegradation was estimated by monitoring changes in mass and mechanical properties (flexural bending) as a function of burial time. The samples were washed with water for the removal of debris on the specimen and finally dried at 60 ± 2 °C temperature for 24 h in air-dried woven before undergoing mass loss and mechanical properties test. Five replicates were evaluated for each type of samples for each test.

The percentage mass loss of treated samples was measured using an electronic balance and determined by Equation (10).

$$\text{Mass Loss } (\%) = \frac{M_i - M_d}{M_i} \times 100 \quad (10)$$

where, M_i is the initial mass and M_d is the mass after the designated burial day. The percentage mass loss was taken from the average of five samples.

Flexural tests for buried samples were conducted according to ISO14125:1998. The deterioration of flexural properties was assessed and the residual flexural property was evaluated using Equation (11).

$$\text{Residual flexural property } (\%) = \frac{P}{P_0} \times 100 \quad (11)$$

where P is the flexural strength or flexural modulus measured after the designated burial day and P_0 is the initial property before burial test.

2.8. Scanning Electron Microscopy (SEM)

A SEM (Philips XL30) was used to study the morphology of the surface degradation of neat PLA, composites and fracture surface of the test specimens. Flax fiber diameter was also measured to obtain the fiber linear density. The surface of the specimens was coated with carbon using an Edward coating system (E306A, USA).

3. Results and Discussion

Figure 2a indicates the produced nonwoven mats of flax/PLA fibers. Figure 2b shows flax/PLA fibers within the mat, confirming the nominally (in-plane) random fiber distribution, which was also preserved in the fabricated PLA molded composite plaques (Figure 2c). The web-forming technique employed here was therefore successful in air-laying nonwoven webs with isotropic fiber orientation distribution.

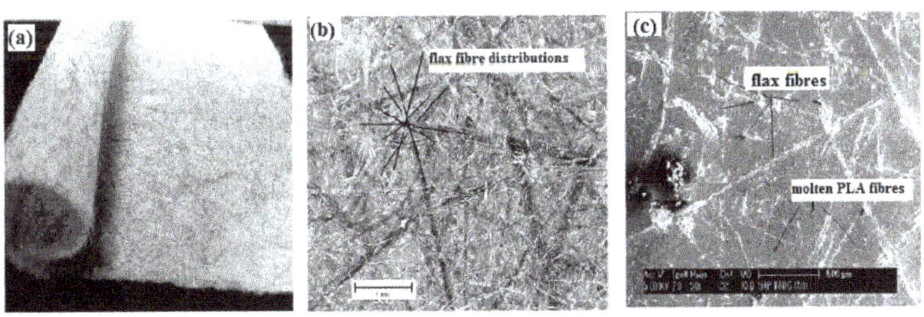

Figure 2. Image of produced flax/PLA nonwoven mat (**a**), flax fiber arrangement within the mat (**b**) and molded composite plaque (**c**).

Matrix dissolution tests on the consolidated NM flax/PLA and UD flax/PLA composites measured a flax fiber mass fraction (W_f) of 48.0% and 49.5%, respectively, similar to the measurements of the unconsolidated preforms. From Table 3 it was found that the calculated density of the nonwoven panel was found to be lower (1.1 g/cm^3) than the aligned panel (1.2 g/cm^3), due to the higher void content (18%) of the nonwoven panel. The latter is indicative of poor impregnation of the flax fibers by the PLA matrix during hot-press molding and suggests that matrix flow was adversely affected by the random orientation of flax fibers in the nonwoven mat composites, despite the use of a higher molding pressure (in comparison to the aligned panel). On the other hand, the calculated porosity content in UD flax/PLA composites was found to be significantly lower (4%). The results indicate that fiber architecture plays an important role on porosity level. The higher void content and greater thickness of the NM flax/PLA composites also led to relatively lower fiber volume fraction (34.3%) compared to that of UD/PLA flax composites (38.6%).

Table 3. Physical and mechanical properties of unreinforced PLA and the composites.

Properties	Neat PLA	Random Flax/PLA	Aligned Flax/PLA
Flax fiber content (wt.%)	0.0	48.0	49.5
Composite density (g/cm^3)	1.24	1.1	1.2
Void content (%)	0.5	18.0	4.0
Tensile strength (MPa)	40.0 ± 3.0	83.0 ± 5.0	151.0 ± 7.0
Tensile modulus (GPa)	3.8 ± 1.0	9.3 ± 1.5	18.5 ± 2.0
Flexural strength (MPa)	82.0 ± 2.0	130.0 ± 5.0	215.0 ± 17.2
Flexural modulus (GPa)	4.0 ± 0.5	9.9 ± 1.0	18.8 ± 1.0
Specific tensile strength (MPa$^{1/2}$. cm^3/g)	5.3	7.7	10.3
Specific tensile modulus (GPa$^{1/3}$. cm^3/g)	1.2	1.8	2.2
Specific flexural strength (MPa$^{1/2}$. cm^3/g)	7.3	10.4	12.2
Specific flexural modulus (GPa$^{1/3}$. cm^3/g)	1.3	2.0	2.2

From Table 3, it was found that the tensile strength of neat PLA was measured to be 40 MPa; this was increased to 83 MPa and 151 MPa with random mat flax reinforcements and unidirectional flax reinforcements, respectively. The tensile modulus of aligned flax composites (18.5 GPa) was also found to be approximately double the random flax composites (9.3 GPa). Similar trends were found in flexural properties, where flexural strength and flexural modulus of UD flax/PLA composites was 60 and 70% higher respectively than NM flax/PLA composites.

Typical tensile stress-strain curves of neat PLA matrix, random flax/PLA composite and aligned flax/PLA composite are shown in Figure 3. In the figure, neat PLA shows a more linear behavior while the composites behave more nonlinearly as the strain increases. The linear phase corresponds to the linear deformation of the fiber and matrix while the nonlinear deformation of the composites has been explained as a three-phase mechanism by Panthapulakkal and Sain [37]. Firstly, the microcrack initiates at the fiber-end/matrix interface and propagates along the fiber lengths; secondly, the matrix undergoes plastic deformation; and finally the microcracks in the matrix open and propagate through the deformed matrix. Due to the pulling out of fibers from the matrix, catastrophic crack propagation also takes place through the matrix. The tensile properties of the biocomposite are higher than neat PLA after the incorporation of flax fiber as anticipated.

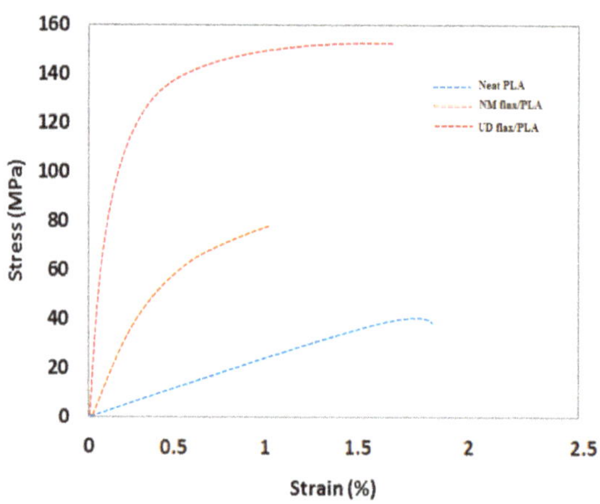

Figure 3. Tensile stress-strain curves of neat PLA and flax/PLA composites.

The significantly better mechanical properties of aligned fiber composites are mainly due to full fiber alignment in the loading direction. It would be expected that properties in the transverse direction of the aligned fiber composites would be comparably lower than those of the isotropic random mat composites. The void content also influences the quality (properties) of the composites. Voids have a negative impact on the mechanical properties of the composites since void can cause weaker bonding. Due to the presence of higher voids in the random fiber composites the mechanical properties were significantly deteriorated.

The fracture surfaces of random fiber composites (Figure 4a), examined through SEM, exhibited substantial fiber debonding and pull-out, which is indicative of poor fiber-matrix adhesion, possibly due to high porosities and dry zones within the composite. Figure 4b presents very little amount of fiber pull-out and lower porosity content specifies good fiber-matrix adhesion in the aligned fiber composites and those fiber properties are extensively transferred to the composite via matrix.

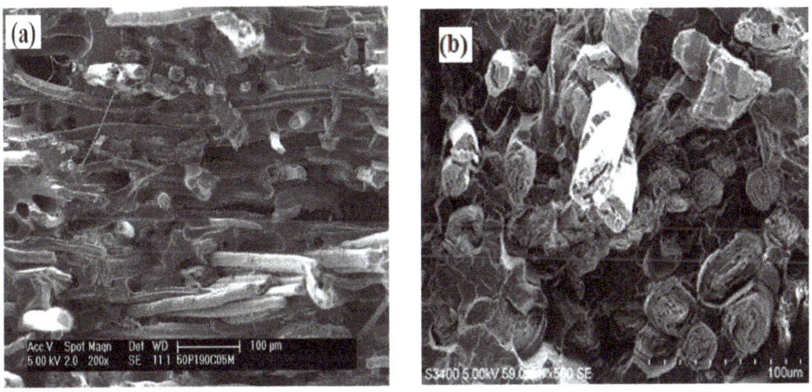

Figure 4. SEM images of tensile fracture surfaces of random fiber oriented (**a**) and unidirectional flax/PLA composites (**b**).

In Table 3, the calculated result of specific flexural strength and specific flexural modulus of both random and aligned panels were given. Both these parameters can be used as material performance indices for the selection of materials in stiffness or strength-critical, light-mass applications, such as an automotive load-floor panel in bending mode. The stiffness-density Ashby chart in Figure 5 compares the performance of the random and aligned flax/PLA composites with other engineering polymers and composites. All the materials that lie on the dashed guide line would give a panel with the same material index parameter, while a panel from materials to the left of the guide line would produce a stiffer panel of the same mass (or lighter panel of same stiffness). It was found that both random mat and aligned flax reinforced PLA thermoplastic panels performed better than wood products but performed worse than conventional glass and carbon fiber polymer composites. It was also revealed that the aligned panel has higher specific strength and stiffness values compared to random mat panel. The specific tensile strength and specific tensile modulus were found to be 10.3 ($MPa^{1/2}$. cm^3/g) and 2.2 ($GPa^{1/3}$. cm^3/g) respectively for UD flax/PLA composites. These values are very much compatible with woven E-glass/Epoxy composites [27]. Therefore, it might be a good choice to replace the woven E-glass/Epoxy composite with this UD flax/PLA panel, where stiffness is more important than strength. Moreover, flax fiber/PLA composite has an extra advantage over glass composite, as it is biodegradable.

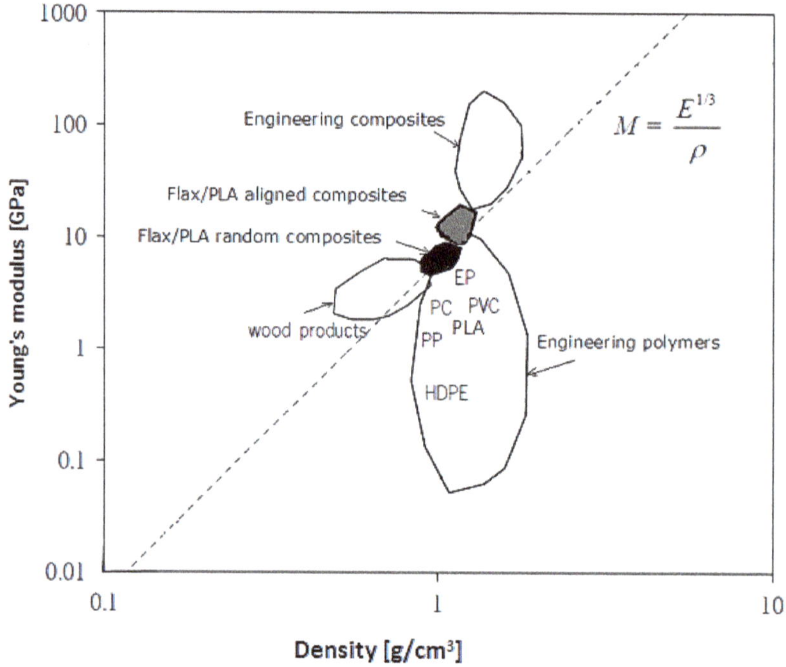

Figure 5. Materials performance chart comparing tensile modulus and density of random and aligned flax/PLA composites with common engineering polymers and composites. The dashed guide line enables materials selection for minimum mass, stiffness-limited design.

Figures 6 and 7 shows the DSC curves of the first and second heating cycle respectively of neat PLA and flax/PLA biocomposites. T_g and T_m were determined from the data of the first heating cycle, as T_g of composite samples was not visible in the second heating cycles, although the T_m differs slightly from the first heating cycle. This is probably due to the relatively slow cooling rate used which led to the crystallization of PLA. The obtained results are summarized in Table 4.

It was observed from Figures 6 and 7 that the glass transition temperature of neat PLA is higher than the values for biocomposites. The introduction of flax fiber decreases the temperature. The lower T_g value indicates the ductility of the composites reduced which may lead to lower impact strength. Non-satisfying impact resistance and low heat distortion temperature can restrict the application of flax/PLA composites in engineering fields [27]. It is also evident that the addition of flax fibers decreases the melting and onset melting temperature and the heat of fusion of the composites marginally because flax fiber acts as a diluent in the PLA matrix.

For UD flax composite, the value of T_g and T_m are higher than NM flax composite. This increase is likely a result of decreasing space available for molecular motion. It was also found that the crystallinity (%) of UD flax/PLA was much higher than NM flax/PLA composites.

In terms of biodegradation assessment of the composite panels, water absorbency performance, mass-loss and residual flexural properties of burial tested samples were investigated.

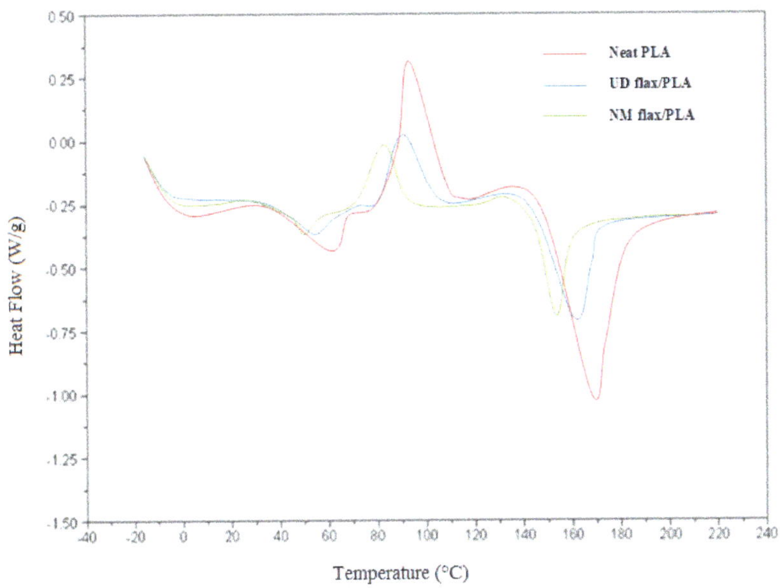

Figure 6. DSC curves of first heating cycle as a function of temperature of neat PLA and flax/PLA biocomposites.

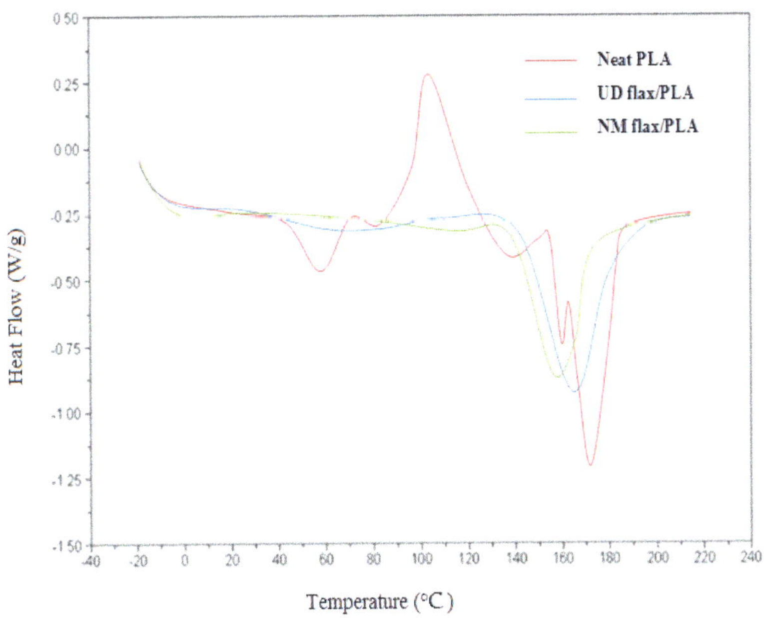

Figure 7. DSC curves of second heating cycle as a function of temperature of neat PLA and flax/PLA biocomposites.

Table 4. Thermal properties of neat PLA and flax/PLA biocomposites.

Sample	Glass Transition Temperature, T_g (°C)	Melting Temperature, T_m (°C)		Melting Enthalpy, ΔH_m (J/g)	Degree of Crystallinity, X_c (%)
		Onset Melting Temperature	Peak Melting Temperature		
Neat PLA	59	160	171	43	46
NM flax/PLA	50	145	155	32	34
UD flax/PLA	54	151	163	39	42

Water absorbency is expressed in terms of mass percentage gain of the composites. It is observed from Figure 8 that, after 60 days soaking period, the NM Flax/PLA composites absorbed water 40% of its mass, which is much higher than that of neat PLA and UD flax/PLA composites. The neat PLA and UD/flax composite absorbed 6 and 11% water respectively at the same time. It was also evident that incorporating flax fibers into the PLA matrix increased water absorption. This can be explained by the higher moisture absorption by the flax fibers. This was also reflected in the measurements of the degree of crystallinity; neat PLA had a crystallinity of 46%, reducing to 34 and 42% for the random and aligned fiber composites, respectively. Flax fibers have a greater amorphous fraction (in the range of 50–70% by dry mass than neat PLA [38] and it is agreed that amorphous regions are more susceptible to moisture absorption and biodegradation [39]. Furthermore, flax fibers composed of almost 70% cellulose and 20% hemicellulose polysaccharides which have a large number of polar hydrophilic –OH groups [40]. In addition, the higher void content of the fiber reinforced composites, in particular, the random mat composites may have a role in accelerating water absorption. In this way, water/moisture initiates the biodegradation process of flax/PLA laminate [30].

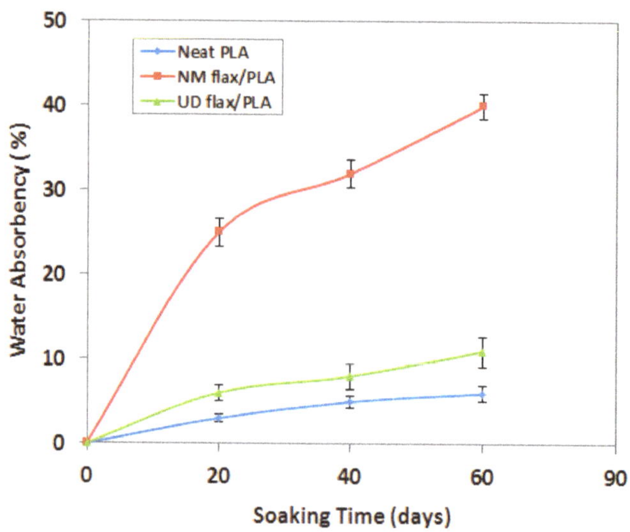

Figure 8. Water absorbency performance of unreinforced and flax reinforced PLA.

Figures 9 and 10 displayed the mass loss (%) of the flax fibers, neat PLA and their composites as a function of burial time. Rapid mass loss (biodegradation) was occurred for flax fibers. From Figure 9 it was found that after 120 days flax fibers lost 90% mass and neat PLA showed minimal (10%) mass loss. Rapid biodegradation action was observed for random mat composites; after 120 days of burial time 27% mass loss was reported while aligned panels exhibited much lower mass loss (19%) over the same period (Figure 10), which was due to lower moisture/water swelling performance of this panel. In Figure 11, SEM images of neat PLA and NM flax/PLA show the evidence of biodegradation, where biodegradation of the matrix started after 60 days. However, SEM images demonstrate that the biodegradation of flax and its composites can start from 30 days in a suitable natural environment (Figure 11f).

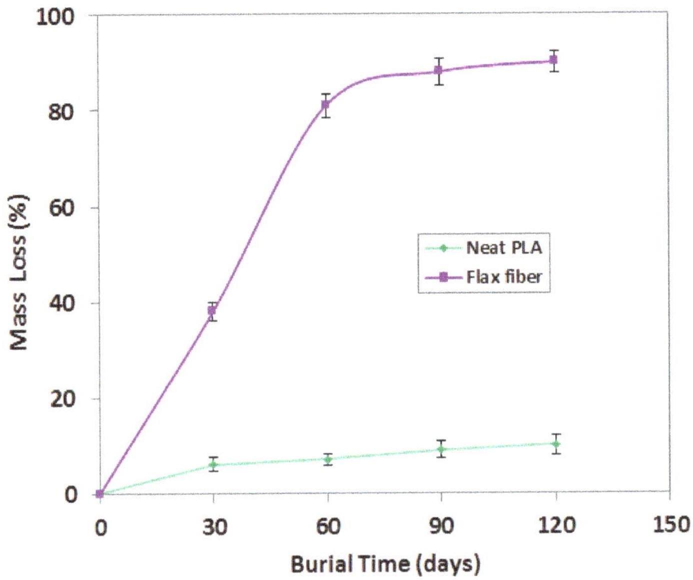

Figure 9. Effect of biodegradations on PLA and flax fibers.

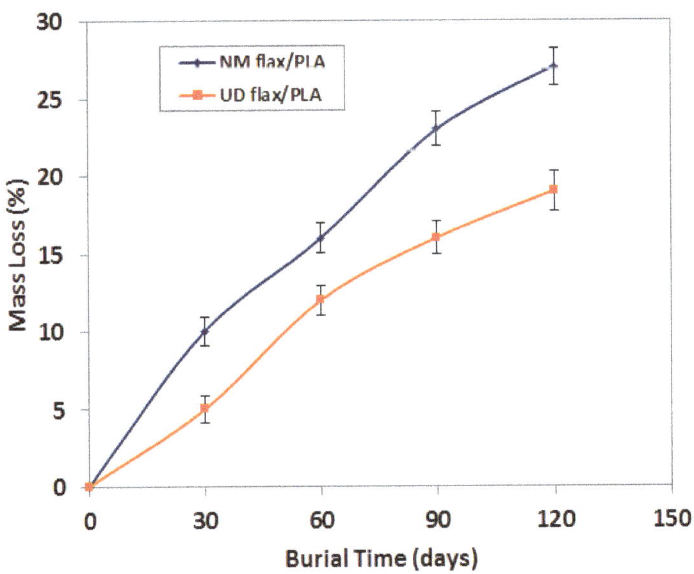

Figure 10. Effect of biodegradations on Flax/PLA composites.

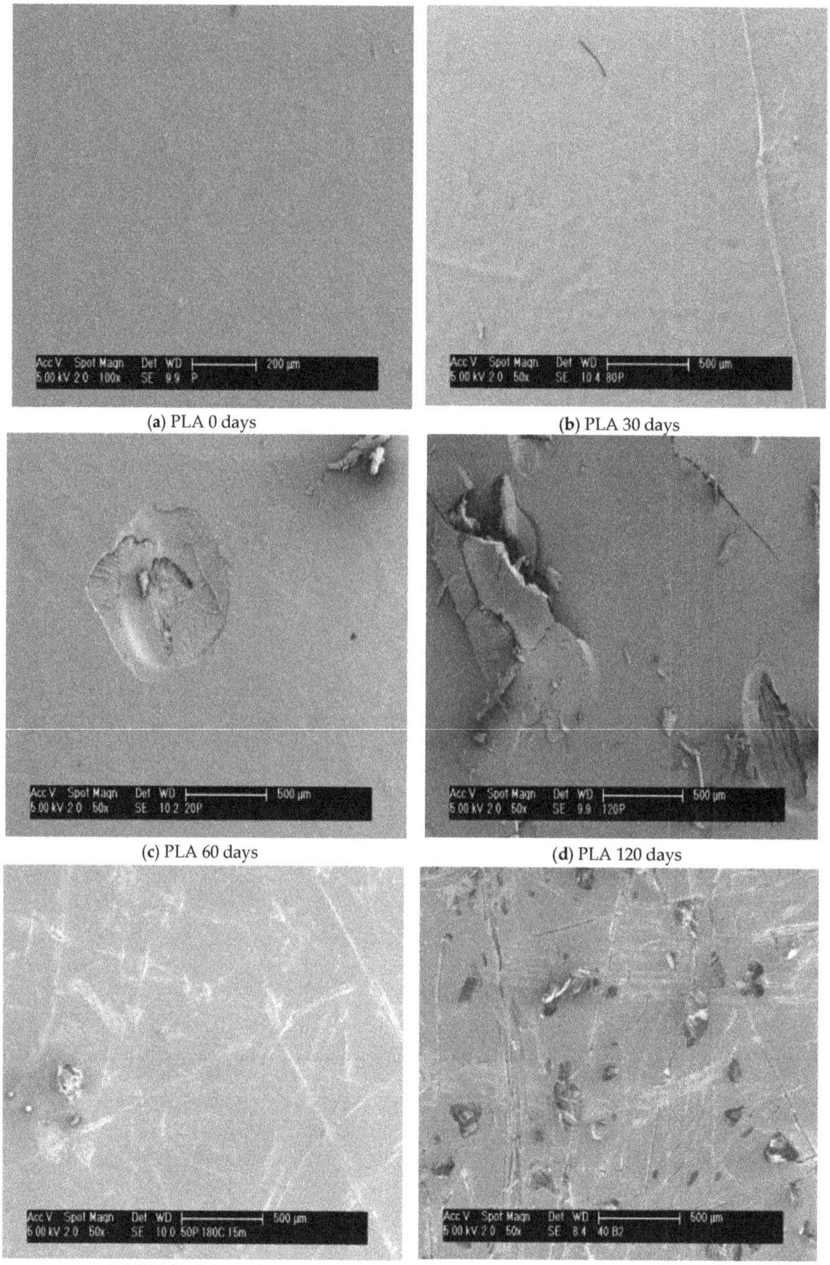

(**a**) PLA 0 days (**b**) PLA 30 days
(**c**) PLA 60 days (**d**) PLA 120 days
(**e**) NM flax/PLA 0 days (**f**) NM flax/PLA 30 days

Figure 11. *Cont.*

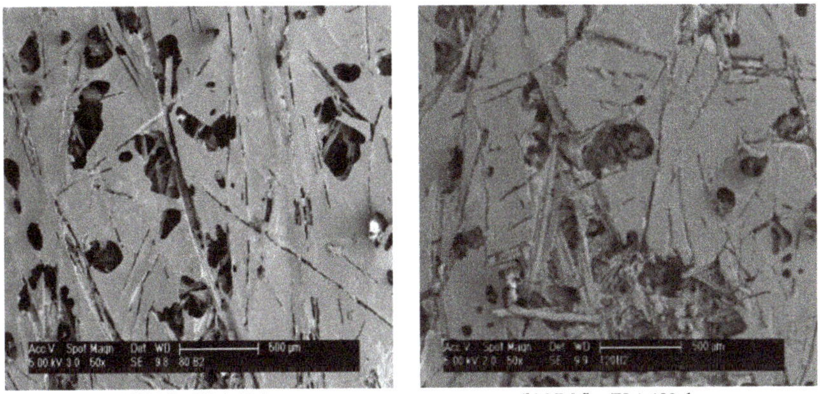

(g) NM flax/PLA 60 days (h) NM flax/PLA 120 days

Figure 11. SEM images of biodegradation effect on neat PLA (**a**–**d**) and NM flax/PLA composites (**e**–**h**).

This biodegradation also triggers a reduction in the mechanical performances of the panels. Figures 12 and 13 indicates that the flexural properties (strength and modulus) of NM flax/PLA composites dropped (by 80%) significantly after 120 days burial test, whereas at that time the flexural strength of neat PLA and UD flax/PLA panels was reduced by 37 and 57%, respectively. The damage of modulus was 20 and 50% for these samples, respectively after 120 days burial time. In all cases, the major loss in properties was within 30 days of burial. The degradation of the properties (particularly stiffness) of random mat panels was substantially greater in comparison to aligned flax panels due to the former's higher water absorbency and subsequent rapid biodegradation, particularly of the flax fiber reinforcement constituent. Due to the presence of more voids in NM flax/PLA panel, it was degraded more after 120 days of burial time which was also observed in Figure 12.

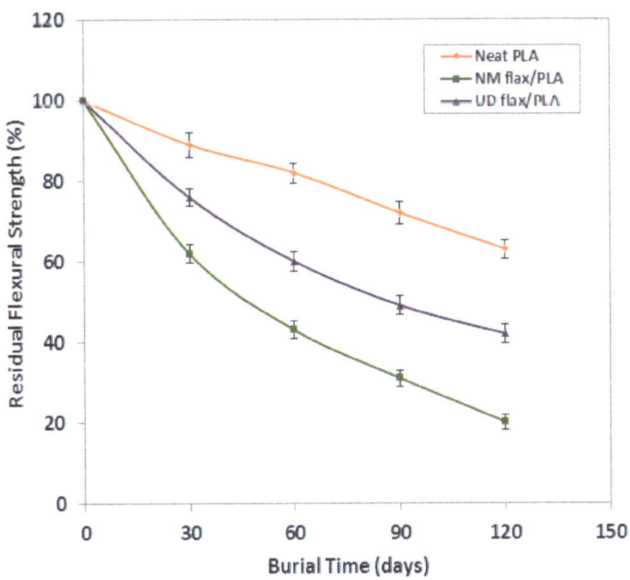

Figure 12. Comparative residual flexural strength of biodegraded composites.

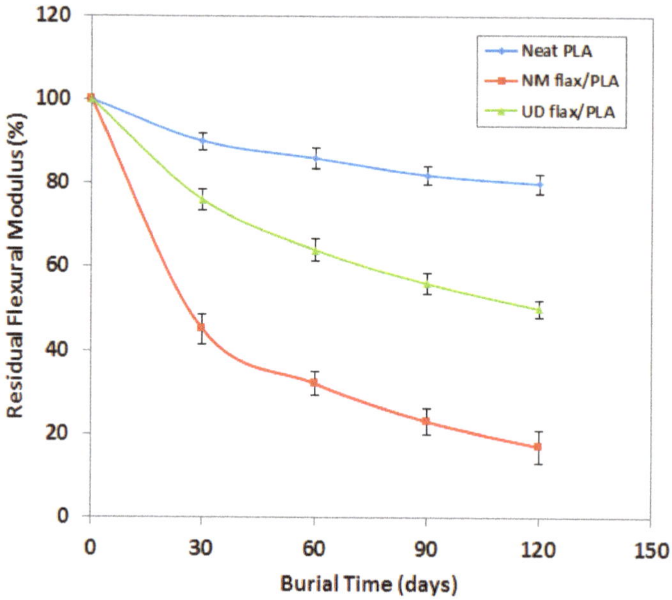

Figure 13. Comparative residual flexural modulus of biodegraded composites.

During soil burial tests, flax fiber is attacked by macro-organisms and micro-organisms. Biodegradation is chiefly introduced by the action of various micro-organisms and the action of micro–organisms are closely associated with the presence of water in soil [41,42]. Macro–organisms may degrade the biocomposites. Flax fiber is composed of cellulose, lignin and pectin which are entirely biodegradable and therefore, the flax fiber is completely biodegradable. On the other hand, the matrix PLA used in this work is also biodegradable. Thus, both the component materials in the manufactured composite are being broken down by micro and macro organisms and enhance the biodegradation. In the case of neat PLA, the biodegradation performance may be attributed to the preferential hydrolysis (breaking up of the polymer into smaller units) [43,44]. The hydrolytic reaction may occur when $-CHCH_3COO$–groups on PLA molecules react with H_2O and produce low molecular mass biocomposites with increased carboxylic groups in the polymer chain (Equation (12)) [43].

$$-CHCH_3COO- + H_2O \rightarrow -CHCH_3OH + HOOC- \quad (12)$$

However, for biocomposites, the biodegradation effect is combined with the fiber-matrix interface degradation. Hydrolytic matrix degradation and fiber debonding may lead to a lower adhesion at the interface and subsequently, to poorer mechanical properties. The influences of the biological environment on the structural change of composites are supported by the SEM image of PLA and NM flax/PLA composites (Figure 11). Nevertheless, the biodegradation effects were delayed for the aligned panel as it was structurally compacted (fibers were well-impregnated by molten matrix) with few voids, therefore, this panel was found more resistant to biodegradation action. If SEM could be done on UD flax/PLA panels then it would be more helpful to understand.

4. Conclusions

In this work, two types of composites, one made of NM flax/PLA and the other one made of UD flax/PLA, were manufactured and their performance was investigated. The results revealed that both types of composites are much stronger than neat PLA as expected. More importantly, the UD flax/PLA showed significantly improved mechanical properties compared to NM flax/PLA. Although both composites consist of same flax fibers (by mass), due to the fiber orientation to the direction of the applied stress, aligned fiber composite provided superior performance under both tensile and bending mode. The tensile and flexural strength of UD flax/PLA composite was increased 82 and 65% respectively and the tensile and flexural modulus was improved 99 and 90% respectively compared to NM flax/PLA composite. The rate of biodegradation of UD flax/PLA structure was also found considerably lower than NM flax/PLA composites due to lower swelling performance. The flexural property of neat PLA, NM flax/PLA and UD flax/PLA are significantly decreased after soil burial. During soil burial test, the UD flax/PLA and NM flax/PLA lost 19 and 27% mass respectively after 120 days observation, the residual flexural strength reduced by 57 and 80% respectively and flexural modulus decreased by 50 and 80% respectively for UD flax/PLA and NM flax/PLA at the same time. The results were confirmed by SEM observations which showed the presence of many large holes and more cracks in the degraded surface of the biocomposites. Significant amount of cavities were found on the surface of the composites which accelerates the biodegradation. Additionally, hydrolysis process may be involved in the degradation of PLA, reducing the properties of the matrix and degrading the interfacial bonding between PLA and flax by molar mass degradation. Specific rigidity values of the composites in bending mode showed that both flax/PLA materials outperform the woven glass fiber reinforced epoxy composite and can replace glass fibers, particularly where stiffness is more important than strength. In terms of environmental impact, biodegradability is a clear advantage of flax/PLA composite than glass fiber reinforced composites as flax/PLA composite is decomposable. Altogether, based on analysis, direct contact with water medium is the most deteriorating environment for biocomposites and therefore underwater applications of these materials are strongly discouraged. However, the results found in the study may be acceptable for the users who are planning to use biodegradable composites in standard atmosphere and for short-term applications in the humid or wet environment.

Author Contributions: M.H.A. and S.A. designed the research work; D.U.S. and A.N.M.M.R. performed all the experiments; M.H.A. and A.N.M.M.R. analyzed the data and prepared the manuscript; all the authors reviewed and edited the final paper.

Funding: The research received no external funding.

Acknowledgments: The authors would like to thank Tilsatec Advanced Textile Materials (UK) and Bangladesh University of Textiles, Bangladesh for supporting this work.

Conflicts of Interest: The authors declare no conflict of interest.

References

1. Mohanty, A.K.; Misra, M.; Drzal, L.T. *Natural Fibers, Biopolymers and Biocomposites*; CRC Press: Boca Raton, FL, USA, 2005.
2. Wallenberger, F.T.; Weston, N. *Natural Fibers, Plastics and Composites*; Springer Science & Business Media: New York, NY, USA, 2003.
3. Awal, A.; Rana, M.; Sain, M. Thermorheological and mechanical properties of cellulose reinforced PLA bio-composites. *Mech. Mater.* **2015**, *80*, 87–95. [CrossRef]
4. Garlotta, D. A literature review of poly (lactic acid). *J. Polym. Environ.* **2001**, *9*, 63–84. [CrossRef]
5. Markarian, J. Biopolymers present new market opportunities for additives in packaging. *Plast. Addit. Compd.* **2008**, *10*, 22–25. [CrossRef]
6. Vroman, I.; Tighzert, L. Biodegradable polymers. *Materials* **2009**, *2*, 307–344. [CrossRef]
7. Yusoff, R.B.; Takagi, H.; Nakagaito, A.N. Tensile and flexural properties of polylactic acid-based hybrid green composites reinforced by kenaf, bamboo and coir fibers. *Ind. Crops Prod.* **2016**, *94*, 562–573. [CrossRef]
8. Mohanty, A.K.; Misra, M.; Drzal, L. Sustainable bio-composites from renewable resources: Opportunities and challenges in the green materials world. *J. Polym. Environ.* **2002**, *10*, 19–26. [CrossRef]
9. Burgueno, R.; Quagliata, M.J.; Mohanty, A.K.; Mehta, G.; Drzal, L.T.; Misra, M. Load-bearing natural fiber composite cellular beams and panels. *Compos. Part A Appl. Sci. Manuf.* **2004**, *35*, 645–656. [CrossRef]

10. Sahoo, S.; Misra, M.; Mohanty, A.K. Enhanced properties of lignin-based biodegradable polymer composites using injection molding process. *Compos. Part A Appl. Sci. Manuf.* **2011**, *42*, 1710–1718. [CrossRef]
11. Cabral, H.; Cisneros, M.; Kenny, J.; Vazquez, A.; Bernal, C. Structure–properties relationship of short jute fiber-reinforced polypropylene composites. *J. Compos. Mater.* **2005**, *39*, 51–65. [CrossRef]
12. Jayaraman, K. Manufacturing sisal–polypropylene composites with minimum fibre degradation. *Comput. Sci. Technol.* **2003**, *63*, 367–374. [CrossRef]
13. Madsen, B.; Hoffmeyer, P.; Thomsen, A.B.; Lilholt, H. Hemp yarn reinforced composites—I. Yarn characteristics. *Compos. Part A Appl. Sci. Manuf.* **2007**, *38*, 2194–2203. [CrossRef]
14. Feng, D.; Caulfield, D.; Sanadi, A. Effect of compatibilizer on the structure-property relationships of kenaf-fiber/polypropylene composites. *Polym. Compos.* **2001**, *22*, 506–517. [CrossRef]
15. Baiardo, M.; Zini, E.; Scandola, M. Flax fibre–polyester composites. *Compos. Part A Appl. Sci. Manuf.* **2004**, *35*, 703–710. [CrossRef]
16. Rubio-López, A.; Artero-Guerrero, J.; Pernas-Sánchez, J.; Santiuste, C. Compression after impact of flax/PLA biodegradable composites. *Polym. Test.* **2017**, *59*, 127–135. [CrossRef]
17. Yan, L.; Chouw, N.; Jayaraman, K. Flax fibre and its composites–A review. *Compos. Part B Eng.* **2014**, *56*, 296–317. [CrossRef]
18. Lucintel, B. *Opportunities in Natural Fiber Composites*; Texas Lucintel: Dallas, TX, USA, 2011.
19. Saravana Bavan, D.; Mohan Kumar, G. Potential use of natural fiber composite materials in India. *J. Reinf. Plast. Compos.* **2010**, *29*, 3600–3613. [CrossRef]
20. Murariu, M.; Dubois, P. PLA composites: From production to properties. *Adv. Drug Del. Rev.* **2016**, *107*, 17–46. [CrossRef]
21. La Mantia, F.; Morreale, M. Green composites: A brief review. *Compos. Part A Appl. Sci. Manuf.* **2011**, *42*, 579–588. [CrossRef]
22. Koronis, G.; Silva, A.; Fontul, M. Green composites: A review of adequate materials for automotive applications. *Compos. Part B Eng.* **2013**, *44*, 120–127. [CrossRef]
23. Foruzanmehr, M.; Vuillaume, P.Y.; Elkoun, S.; Robert, M. Physical and mechanical properties of PLA composites reinforced by TiO_2 grafted flax fibers. *Mater. Des.* **2016**, *106*, 295–304. [CrossRef]
24. Lim, L.-T.; Auras, R.; Rubino, M. Processing technologies for poly (lactic acid). *Prog. Polym. Sci.* **2008**, *33*, 820–852. [CrossRef]
25. Ma, H.; Joo, C.W. Structure and mechanical properties of jute—Polylactic acid biodegradable composites. *J. Compos. Mater.* **2011**, *45*, 1451–1460.
26. Shanks, R.; Hodzic, A.; Ridderhof, D. Composites of poly (lactic acid) with flax fibers modified by interstitial polymerization. *J. Appl. Polym. Sci.* **2006**, *101*, 3620–3629. [CrossRef]
27. Couture, A.; Lebrun, G.; Laperrière, L. Mechanical properties of polylactic acid (PLA) composites reinforced with unidirectional flax and flax-paper layers. *Compos. Struct.* **2016**, *154*, 286–295. [CrossRef]
28. Alimuzzaman, S.; Gong, R.H.; Akonda, M. *Impact Property of PLA/flax Nonwoven Biocomposite*; Conference Papers in Materials Science; Hindawi: New York, NY, USA, 2013.
29. Lee, S.H.; Kang, T.J. Mechanical and impact properties of needle punched nonwoven composites. *J. Compos. Mater.* **2000**, *34*, 816–840. [CrossRef]
30. Alimuzzaman, S.; Gong, R.H.; Akonda, M. Nonwoven polylactic acid and flax biocomposites. *Polym. Compos.* **2013**, *34*, 1611–1619. [CrossRef]
31. Hearle, J.; Stevenson, P. Nonwoven fabric studies: Part III: The anisotropy of nonwoven fabrics. *Text. Res. J.* **1963**, *33*, 877–888. [CrossRef]
32. Krenchel, H. *Fibre Reinforcement; Theoretical and Practical Investigations of the Elasticity and Strength of Fibre-Reinforced Materials*; Akademisk Forlag: Copenhagen, Denmark, 1964.
33. Bledzki, A.K.; Jaszkiewicz, A.; Scherzer, D. Mechanical properties of PLA composites with man-made cellulose and abaca fibres. *Compos. Part A Appl. Sci. Manuf.* **2009**, *40*, 404–412. [CrossRef]
34. Kim, H.-S.; Kim, H.-J.; Lee, J.-W.; Choi, I.-G. Biodegradability of bio-flour filled biodegradable poly (butylene succinate) bio-composites in natural and compost soil. *Polym. Dégrad. Stab.* **2006**, *91*, 1117–1127. [CrossRef]
35. Kim, H.S.; Yang, H.S.; Kim, H.J. Biodegradability and mechanical properties of agro-flour–filled polybutylene succinate biocomposites. *J. Appl. Polym. Sci.* **2005**, *97*, 1513–1521. [CrossRef]
36. Alimuzzaman, S.; Gong, R.H.; Akonda, M. Biodegradability of nonwoven flax fiber reinforced polylactic acid biocomposites. *Polym. Compos.* **2014**, *35*, 2094–2102. [CrossRef]

37. Panthapulakkal, S.; Sain, M. Injection-molded short hemp fiber/glass fiber-reinforced polypropylene hybrid composites—Mechanical, water absorption and thermal properties. *J. Appl. Polym. Sci.* **2007**, *103*, 2432–2441. [CrossRef]
38. Satyanarayana, K.G.; Arizaga, G.G.; Wypych, F. Biodegradable composites based on lignocellulosic fibers—An overview. *Prog. Polym. Sci.* **2009**, *34*, 982–1021. [CrossRef]
39. Plackett, D.; Andersen, T.L.; Pedersen, W.B.; Nielsen, L. Biodegradable composites based on L-polylactide and jute fibres. *Compos. Sci. Technol.* **2003**, *63*, 1287–1296. [CrossRef]
40. John, M.J.; Anandjiwala, R.D. Recent developments in chemical modification and characterization of natural fiber-reinforced composites. *Polym. Compos.* **2008**, *29*, 187–207. [CrossRef]
41. Rudeekit, Y.; Numnoi, J.; Tajan, M.; Chaiwutthinan, P.; Leejarkpai, T. Determining biodegradability of polylactic acid under different environments. *J. Met. Mater. Miner.* **2008**, *18*, 83–87.
42. Kumar, R.; Yakubu, M.; Anandjiwala, R. Biodegradation of flax fiber reinforced poly lactic acid. *Express Polym. Lett.* **2010**, *4*, 423–430. [CrossRef]
43. Mohammad Khanlou, H.; Hall, W.; Woodfield, P.; Summerscales, J.; Francucci, G. The mechanical properties of flax fibre reinforced poly (lactic acid) bio-composites exposed to wet, freezing and humid environments. *J. Compos. Mater.* **2018**, *52*, 835–850. [CrossRef]
44. Bayerl, T.; Geith, M.; Somashekar, A.A.; Bhattacharyya, D. Influence of fibre architecture on the biodegradability of FLAX/PLA composites. *Int. Biodeterior. Biodégrad.* **2014**, *96*, 18–25. [CrossRef]

fibers

MDPI

Article

Heat and Mass Transfer Properties of Sugar Maple Wood Treated by the Thermo-Hygro-Mechanical Densification Process

Qilan Fu [1], Alain Cloutier [1,*] and Aziz Laghdir [2]

[1] Center de Recherche sur les Matériaux Renouvelables (Renewable Materials Research Center), Département des Sciences du Bois et de la Forêt (Department of Wood and Forest Science), Université Laval, Québec, QC G1V 0A6, Canada; qilan.fu.1@ulaval.ca

[2] Research Scientist, Service de Recherche et D'expertise en Transformation des Produits Forestiers (Research and Expertise Service on Transformation of Forest Products), Amqui, QC G5J 1K3, Canada; aziz.laghdir@serex.qc.ca

* Correspondence: alain.cloutier@sbf.ulaval.ca; Tel.: +1-418-656-5851

Received: 29 June 2018; Accepted: 17 July 2018; Published: 24 July 2018

Abstract: This study investigated the evolution of the density, gas permeability, and thermal conductivity of sugar maple wood during the thermo-hygro-mechanical densification process. The results suggested that the oven-dry average density of densified samples was significantly higher than that of the control samples. However, the oven-dry density did not show a linear increase with the decrease of wood samples thickness. The radial intrinsic gas permeability of the control samples was 5 to 40 times higher than that of densified samples, which indicated that the void volume of wood was reduced notably after the densification process. The thermal conductivity increased by 0.5–1.5 percent for an increase of one percent moisture content for densified samples. The thermal conductivity of densified wood was lower than that of the control samples. The densification time had significant effects on the oven-dry density and gas permeability. Both densification time and moisture content had significant effects on thermal conductivity but their interaction effect was not significant.

Keywords: density; gas permeability; thermal conductivity; densification

1. Introduction

The heat and mass transfer properties of wood or wood-based materials are essential characteristics required for a variety of purposes, including the heat and mass transfer modeling during the densification process and characterization of densified wood as a building material. Among others, typical properties required are density, permeability, and thermal conductivity. A proper characterization of these properties is crucial for increasing the accuracy of model predictions. However, during the thermo-hygro-mechanical (THM) densification process, density, permeability, and thermal conductivity of wood are all time-dependent, which makes the characterization of these parameters difficult.

Bulk flow is the principal mechanism for the transport of fluids through wood, which occurs through the voids of the wood under a static or capillary pressure gradient [1]. The bulk flow rate of fluids is determined by wood permeability. During the hot-pressing process of wood composite materials, the gas permeability controls the convective heat transfer from surface layers to the core layer and impacts the movement of the vapor from the core to the edges [2]. Since gas permeability depends largely on the pore structure of the fiber or particle mat, the densification treatment should have a direct effect on the permeability. Comstock [3] reported that the arrangement of wood principal directions has more impact on its gas permeability than any other parameter. In some species, the longitudinal permeability could be 10^6 higher than in the transverse direction, due to the arrangement of the wood cells.

A few researchers investigated the gas permeability of wood based panels using experimental methods [4–8]. Almost all the methods described in the literature are based on measuring the amount of gas flow at a given pressure gradient applied across the sample. Defo et al. [9] measured radial and tangential gas permeability values of sugar maple wood at 12% of moisture content (MC) (oven-dry density varied from 587 to 676 kg/m^3) between 2.04×10^{-17} m^3_{air} m^{-1}_{wood} and 2.84×10^{-17} m^3_{air} m^{-1}_{wood}. Von Haas et al. [6] reported that the MC had almost no effect on gas permeability for low density samples and a slight effect on samples with a density above 900 kg/m^3. Moreover, air permeability was at least two orders of magnitude higher than steam permeability, which might be due to the swelling of wood and the viscosity of the fluid on the superficial permeability [4,10].

Thermal conductivity is an important material property in determining the heat transfer rate [11]. The thermal conductivity of wood is affected by several basic factors: density, temperature, MC, extractive content, grain direction, structural irregularities, such as checks, and knots and microfibril angle, among which density and MC are predominant [12]. Troppová et al. [13] found that higher temperatures resulted in larger differences between the thermal conductivity values of wood-based fiberboards in the oven dry condition and at 14.2% MC. Thermal conductivity in the radial direction was reported to be about 5% to 10% higher than in the tangential direction [14]. Thermal conductivity along the grain has been reported to be 1.5 to 2.8 times higher than across the grain but the reported values vary widely. For example, Maclean [15] found that the thermal conductivity along the longitudinal direction is about 2.25 to 2.75 times higher than transverse thermal conductivity with an average of approximately 2.5.

The steady-state method is normally applied to measure the thermal conductivity of wood [16]. A large number of empirical equations could be found in the literature, to describe the relationship between wood thermal conductivity, density, and MC [1,10,15,17]. However, these empirical equations are applicable only within a limited range of MC and density levels. Hence, these relations may not be appropriate to describe the variation of the thermal conductivity of wood undergoing THM densification because both MC and density vary continuously during the process. In addition, little empirical data were found for the thermal conductivity of sugar maple wood at different density levels. Therefore, the objective of the current study was to investigate the variation of density, gas permeability, and thermal conductivity of sugar maple wood during the THM densification process.

2. Materials and Methods

Sugar maple (*Acer saccharum* Marsh) wood was selected for this study. This species is a diffuse-porous hardwood normally used for in-door applications such as flooring and furniture [18]. Thin sawn strips of sugar maple wood were provided by a hardwood flooring plant (Lauzon, Distinctive Hardwood Flooring Inc., Papineauville, QC, Canada). Their average apparent density (20 °C and 65% relative humidity (RH)) was 734 (standard deviation: 8.4) kg/m^3 and their dimensions were 5.7 mm (radial) × 84.0 mm (tangential) × 695.0 mm (longitudinal). When they were received, the strips were stored in a conditioning room at 20 °C and 65% RH until an equilibrium moisture content of approximately 12% was achieved. Ten groups of 8 strips were densified for 0 min (control sample), 5 min, 10 min, 15 min, 20 min, 25 min, 30 min, 35 min, 40 min, and 45 min, respectively. The strips were used for experimental determination of the evolution with time of density, gas permeability, and thermal conductivity during the THM densification process.

2.1. Thermo-Hygro-Mechanical Densification Process

A steam injection press (Dieffenbacher, Alpharetta, GA, USA) with dimensions of 862 mm × 862 mm was used for the densification treatment (Figure 1). Steam injection holes with a diameter of 1.5 mm were distributed uniformly at 32 mm intervals on both the top and bottom platens of the press. To reduce wood surface carbonization and distribute the steam uniformly, both surfaces of the specimens were covered with a thin heat-resistant fabric permeable to steam made of Nomex® III A manufactured by Dupont™ [19]. The two press platens were preheated to the target temperature (200 °C) before treatment. The upper platen reached the specimens within 86 s during press closing.

Figure 1. Steam injection hot press used for THM densification treatment.

For all of the treatments except for the control group, the densification process was pre-set in the computer control system. Steam was continuously injected during the whole process at a maximum manometer pressure of 550 kPa, while the specimens were pressed under an increasing mechanical manometer platen pressure up to 6 MPa [20]. The evolution of steam pressure and platen pressure during the whole process is presented in Figure 2 [20]. The whole densification process was divided into ten steps according to the treatment time (0 min, 5 min, 10 min, 15 min, 20 min, 25 min, 30 min, 35 min, 40 min, and 45 min). The density, gas permeability, and thermal conductivity of all samples were determined for each treatment time in order to track their variation during the THM densification process. For the samples densified for 5 min, the control system stopped the process 5 min after the two hot platens closed, even though the maximum platen pressure had not been reached [20]. For the other treatments, the control system stopped the process after 10 min, 15 min, 20 min, 25 min, 30 min, 35 min, 40 min, 45 min respectively. In these cases, both the maximum steam pressure and platen pressure were reached. All the treated samples were then stored in a conditioning room at 20 °C and 65% RH until their equilibrium moisture content was reached prior to their properties determination.

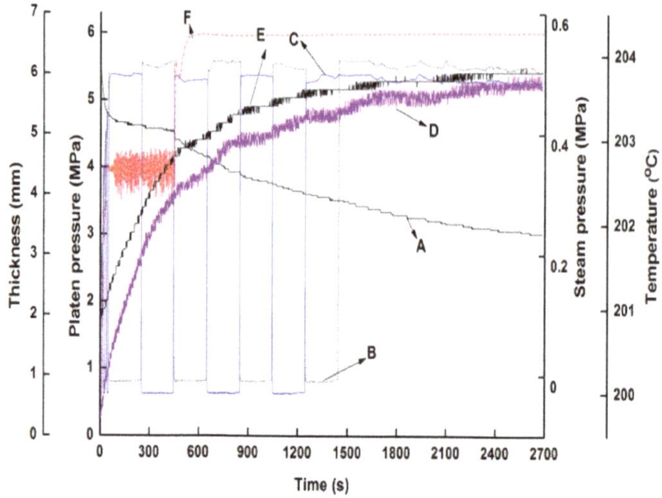

Figure 2. Thermo-hygro-mechanical densification process. **A**: Thickness; **B**, **C**: top and bottom platen steam pressure, respectively; **D**, **E**: top and bottom platen temperature; **F**: platen load.

2.2. *Properties Determination—Oven-Dry Average Density*

Three specimens for each densification time with dimensions 50 mm × 50 mm were oven-dried and used to measure the density using an X-ray densitometer (Quintek Measurements Systems model QDP-01X, Knoxville, TN, USA) at intervals of 0.02 mm through the thickness. The average value (n = 3) was used as the final oven-dry density of each group.

2.3. *Permeability Measurement*

Three discs of 50 mm in diameter for each densification time were prepared for the gas permeability measurement. A special device developed in our laboratory by Lihra et al. [7] was used to measure the transverse gas permeability of the wood samples. The gas permeability was measured in this study with air using the apparatus shown in Figure 3. A cylinder of compressed air equipped with a pressure regulator was used to regulate the flow of gas at the desired pressure. In addition, a silicon seal was used on the edge of each disc in order to make a tight seal with a rubber sleeve surrounding it. A pressure of 600 kPa was applied to the rubber sleeve to prevent air leaks through the specimen edge. Two basswood discs (*Tilia americana*) with high longitudinal gas permeability were placed both in the inlet and outlet sides of the specimen to distribute the air flow [8]. Five flowmeters (Figure 3) with increasing range were installed to measure the gas flow rate through the samples. For each measurement, the flowmeter with a larger range (flowmeter 5) was firstly used. If there was no value provided, it was closed and the next one was used. This procedure was repeated until the gas flow rate could be measured. Each disc (n = 3 for each group) was measured at four pressure levels (ΔP—values measured from pressure gage B): 200 kPa, 250 kPa, 300 kPa, and 350 kPa, respectively.

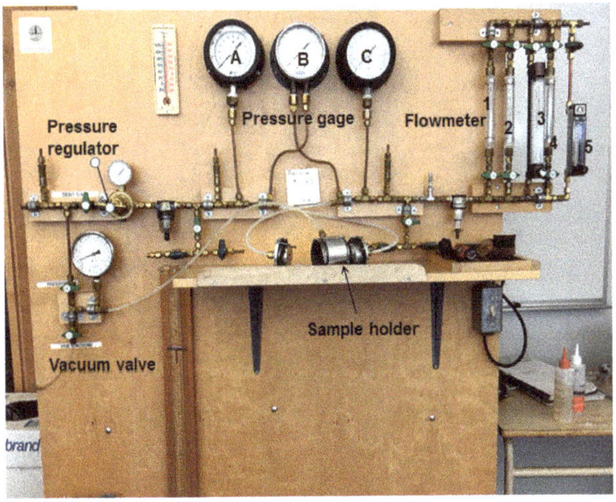

Figure 3. Apparatus used for gas permeability measurements.

The steady-state gas flow through wood can be characterized by Darcy's law. It could be stated as:

$$k_p{}^* = \frac{Q_g L}{A_s \Delta P}\frac{P}{\overline{P}} \quad (1)$$

where $k_p{}^*$ is apparent gas permeability with slip flow ($m^3_{gas}\ m^{-1}_{wood}\ s^{-1}\ Pa^{-1}$), Q_g is the volumetric gas flow rate (m^3/s), L is the length in the flow directioncorresponding to the thickness of the sample (m), A_s is gas flow area (m^2), ΔP is the pressure differential between the inlet side and outlet side (Pa) ($\Delta P = P1 - P2$), P1 is the inlet air pressure (Pa), P2 is the outlet air pressure (Pa), P is the pressure at which Q_g was measured (Pa), $\overline{P}$ is the arithmetic average pressure (Pa), $\overline{P} = \frac{(P_1+P_2)}{2}$.

The apparent gas permeability $k_p{}^*$ from Equation (1) includes Knudsen diffusion, also called slip flow. When a gas flows through a capillary whose diameter is in the same order of magnitude as the average free path between the gas molecules, slip flow becomes significant and must be considered in the permeability measurement. The gas permeability corrected for slip flow could be obtained from the Klinkenberg equation [21]:

$$k_p{}^* = k_p \times s \quad (2)$$

$$s = 1 + \frac{3.8\lambda}{r} \quad (3)$$

$$\lambda = \frac{2\mu}{\overline{P}}\sqrt{\frac{RT}{M_a}} \quad (4)$$

$$K = k_p \times \mu \quad (5)$$

where k_p is the apparent gas permeability corrected for slip flow ($m^3_{gas}\ m^{-1}_{wood}\ s^{-1}\ Pa^{-1}Pa$), s is the slip flow factor, λ is the average free path between gas molecules (m), r is the diameter of the capillary, R is the universal gas constant (8.31 J/mol/K), T is the absolute temperature (K), M_a is the molecular weight of air (kg/mol). K is the intrinsic gas permeability ($m^3_{gas}\ m^{-1}_{wood}$), μ is viscosity of fluid (Pa·s) (for air at room temperature $\mu = 1.845 \times 10^{-5}$ Pa·s). k_p represents the "true" gas permeability corrected for slip flow and can be determined graphically from the intercept of a plot of $k_p{}^*$ against the reciprocal average pressure ($1/\overline{P}$) [7,21].

2.4. Thermal Conductivity Measurement

Four specimens for each densification time with dimensions 152.4 mm × 170.0 mm were prepared for thermal conductivity measurement using the apparatus LaserComp Fox 314 (TA instruments, New Castle, DE, USA) shown in Figure 4a. The Fox 314 instrument was designed according to the ASTM C518-04 standard test method for steady-state thermal transmission properties by means of the heat flow meter apparatus. The specimen was placed between two heating plates (Figure 4b) with different temperature for a sufficient length of time to obtain a uniform temperature gradient throughout the sample. The temperature of the upper heating plate was set at 10 °C and that of the base heating plate was set at 35 °C. The temperature equilibrium of the system is considered to be reached when the temperatures of the two plates are stable within ±0.2 °C after the set point has been reached. During the test, the auto thickness mode was selected, and the sample's thickness was determined automatically by the instrument's digital thickness measurement system.

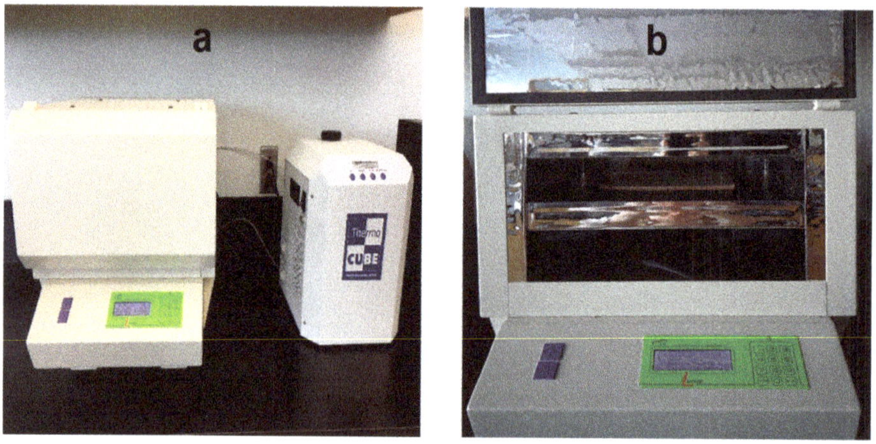

Figure 4. (**a**) Thermal conductivity testing apparatus; (**b**) Position of the sample and two heating plates.

The thermal conductivity is significantly affected by MC. To investigate the influence of MC on the thermal conductivity, each specimen (n = 4 for each group) was measured at three moisture content levels (0%, 6%, and 12%), respectively. Finally, the thermal conductivity was described as a function of densification time and MC.

2.5. Statistical Analysis

An analysis of variance (ANOVA) was performed to investigate the effect of densification time on the oven-dry density and gas permeability, and the effects of the densification time and MC on the thermal conductivity of densified sugar maple wood using SAS 9.4 (SAS Institute Inc., Cary, NC, USA) at significance level $\alpha = 0.05$. Duncan's test was conducted for multiple comparisons between average values obtained under different treatments.

3. Results

3.1. Density Profile and Oven-Dry Density

Typical density profiles of the control sample (densification time = 0 min) and samples densified for 10 min and 40 min, respectively, were selected to investigate the density distribution of sample through its thickness undergoing different densification time treatments. As shown in Figure 5, the density of the control sample was almost constant throughout the thickness, with the exception of the lower density values observed on both surfaces. The samples densified for 10 min showed a higher density in the core than at the surface, this might have been caused by the large spring back after the press opening. In our previous research [22], the density profile of the samples densified at lower temperatures (180 °C and 190 °C) without steam showed similar tendency. This result also suggested that densification for 10 min was not enough, as the compressed deformation was not stable. The compressed sample surface experienced large springback, resulting in a non-homogeneous density distribution. The density was more homogeneous in the core for the samples densified for 40 min, and a higher density at the surface than in the core was found. These observations were the same for samples densified at 200 °C with steam [22]. The heat distribution across the transverse direction was likely more homogeneous for samples densified for longer time with steam.

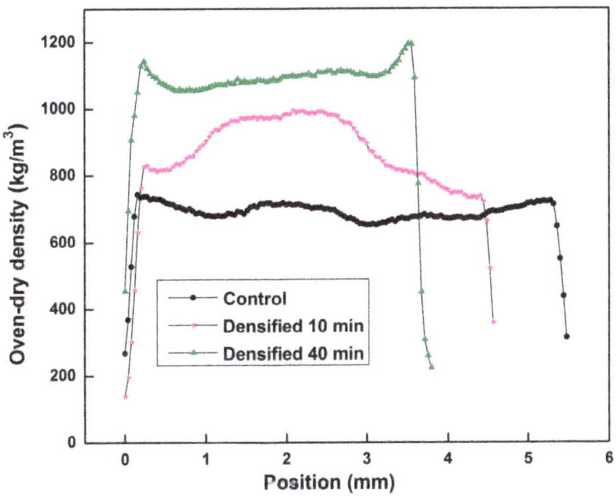

Figure 5. Typical density profile of wood samples densified at different time.

Analysis of variance results of average oven-dry density versus densification time are presented in Table 1. The results show that the densification time had a significant effect on the oven-dry density ($p = 0.0002$). Table 2 presents the evolution of the oven-dry density and thickness of the samples during the THM densification process. It can be seen that the thickness of the samples decreases in general with densification time. Most of the thickness reduction occurred within the first 5–15 min. This main reduction of thickness may be caused by the decrease of the void volume, and consequently resulted in an increase of the oven-dry density. However, the oven-dry density did not present a linear increase with the decrease of thickness. When the densification time is over 40 min, the oven-dry density decreased, which might be due to the degradation of wood components for longer densification time at 200 °C, resulting in a decrease in the overall density [19].

Table 1. Analysis of variance results of oven-dry density versus densification time.

Source	Sum of Squares	Degrees of Freedom	Mean Square	*F* Value	*p* Value	Remarks
Densification time	280,224.4	9	31,136.0	6.8	0.0002	Significant

Table 2. Oven-dry density of the control and samples densified at different time.

Treatment Time (min)	Thickness (mm) (n = 3)	Oven-Dry Density (kg/m^3) (n = 3)
0	5.63 (0.11)	694.3 (4.1) [c,*]
5	4.49 (0.15)	848.6 (82.8) [b]
10	4.26 (0.04)	845.4 (76.1) [b]
15	3.81 (0.09)	917.3 (61.0) [a,b]
20	3.8 (0.10)	948.3 (106.2) [a,b]
25	3.63 (0.04)	947.1 (43.5) [a,b]
30	3.6 (0.07)	952.7 (81.9) [a,b]
35	3.68 (0.06)	1039.6 (60.6) [a]
40	3.57 (0.09)	1031.6 (44.2) [a]
45	3.51 (0.08)	969.5 (65.6) [a,b]

Values in parenthesis are standard deviations; * Duncan's test results, average values with the same letter indicate no significant difference at $\alpha = 0.05$.

3.2. Gas Permeability

A plot of $k_p{}^*$ against $1/\overline{P}$ was made for each sample to correct the effect of slip flow. Figure 6 presents a typical relationship of $k_p{}^*$ against $1/\overline{P}$ for radial flow in a control sample. A linear relationship can be observed in Figure 6 with a small but significant slope indicating the occurrence of slip flow. Therefore, the measured values of $k_p{}^*$ should be corrected for slip flow to obtain the gas permeability k_p from the intercept of the plot of $k_p{}^*$ against $1/\overline{P}$ [7,21].

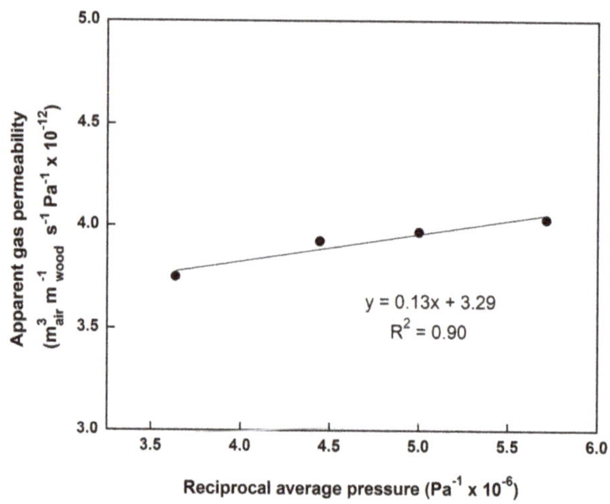

Figure 6. Typical relationship of the apparent gas permeability against the reciprocal average pressure for radial flow of the control sample.

The evolution of radial intrinsic gas permeability during the THM densification process is presented in Figure 7. Analysis of variance results of intrinsic gas permeability versus densification time are presented in Table 3. Table 4 presents the intrinsic gas permeability values of the control and densified samples, and the results of Duncan's test for all of the treatments.

The radial intrinsic gas permeability of the control sugar maple wood sample with an average oven-dry density of 694.3 kg/m^3 is 5.93×10^{-17} m^3_{air} m^{-1}_{wood}. As shown in Figure 7 and Table 4, the intrinsic gas permeability of wood decreased rapidly at the beginning of the densification treatment compared with that of the control sample. However, no statistical difference was found between the intrinsic gas permeability of samples densified at 5, 10, and 15 min, which might be due to their densities which were not significantly different. From 20 to 45 min, the intrinsic gas permeability decreased with increasing of densification time. The intrinsic gas permeability of the control sample could be 5 to 40 times higher than for densified samples, which suggested that the void volume of wood reduced notably after densification. This might be due to the large difference between the radial intrinsic gas permeability of the control and densified samples, the whole data did not meet the assumption of a normal distribution. After performing normality test using SAS, it was found that a logarithm transformation (log_{10}) was needed. Table 3 revealed that the densification time had also a significant effect on wood permeability ($p < 0.0001$).

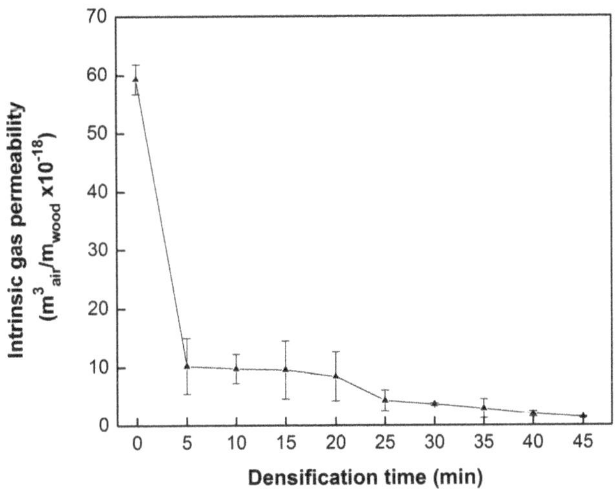

Figure 7. Evolution of the radial intrinsic gas permeability during the densification process.

Table 3. Analysis of variance results of the radial intrinsic gas permeability versus densification time.

Source	Sum of Squares	Degrees of Freedom	Mean Square	F Value	p Value	Remarks
Densification time (data after logarithmic transformation treatment)	5.9	9	0.66	22.61	<0.0001	Significant

Table 4. Radial intrinsic gas permeability of the control and samples densified for different times.

Treatment Time (min)	Oven-Dry Density (kg/m^3) (n = 3)	Intrinsic Gas Permeability $\left(m^3_{air}\ m^{-1}_{wood}\right)$ (n = 3)	Intrinsic Gas Permeability (after Logarithm Transformation)
0	694.3 (4.1) ^{c,*}	5.93×10^{-17} (2.57×10^{-18})	-16.23 ^{a,*}
5	848.6 (82.8) ^{b}	1.02×10^{-17} (4.71×10^{-18})	-17.02 ^{b}
10	845.4 (76.1) ^{b}	9.74×10^{-18} (2.48×10^{-18})	-17.02 ^{b}
15	917.3 (61.0) ^{a,b}	9.53×10^{-18} (4.95×10^{-18})	-17.06 ^{b}
20	948.3 (106.2) ^{a,b}	8.40×10^{-18} (4.19×10^{-18})	-17.12 ^{b,c}
25	947.1 (43.5) ^{a,b}	4.28×10^{-18} (1.76×10^{-18})	-17.40 ^{c,d}
30	952.7 (81.9) ^{a,b}	3.63×10^{-18} (2.00×10^{-19})	-17.44 ^{d,e}
35	1039.6 (60.6) ^{a}	2.87×10^{-18} (1.64×10^{-18})	-17.58 ^{d,e,f}
40	1031.6 (44.2) ^{a}	1.95×10^{-18} (4.8×10^{-19})	-17.72 ^{e,f}
45	969.5 (65.6) ^{a,b}	1.4×10^{-18} (0.84×10^{-19})	-17.86 ^{f}

Values in parenthesis are standard deviations; * Duncan's test results, average values with the same letter indicate no significant difference at $\alpha = 0.05$.

3.3. Thermal Conductivity

Table 5 shows the analysis of variance results of thermal conductivity versus densification time and MC. The thermal conductivity values of the control samples and densified samples are presented in Table 6. Table 5 reveals that both densification time and MC had significant effect on the thermal conductivity, but the interaction between the densification time and MC was not significant.

Table 5. Analysis of variance results of thermal conductivity versus densification time and moisture content.

Source	Sum of Squares	Degrees of Freedom	Mean Square	*F* Value	*p* Value	Remarks
Densification time	0.014	9	0.002	38.88	<0.0001	Significant
Moisture content	0.001	2	0.001	19.18	<0.0001	Significant
Densification time * Moisture content	0.0002	18	0	0.24	0.99	

* The interaction between densification time and moisture content.

There is a common agreement that MC has an important effect on wood thermal conductivity. As shown in Table 6, for all of treatment times, thermal conductivity increased with increasing MC. It could be found that the thermal conductivity increased by 0.5–1.5 percent for an increase of one percent MC for densified sugar maple wood. In particular, it can be noticed that the thermal conductivity of densified samples is slightly lower than that of the control samples. This was not expected given that wood thermal conductivity generally increases with increasing density [1,12]. However, according to ThermoWood Handbook [23], the thermal conductivity of heat treated wood (230 °C, 3–5 h) is reduced by 20–25% compared with normal untreated softwoods (pine and spruce). The underlying reasons for the decrease in thermal conductivity after densification treatment are not entirely clear. It might be due to the alteration of the crystalline structure of cellulose chains at higher treatment temperature, resulting in strength loss and changes in its ability to conduct heat at the cellular level [16].

Table 6. Thermal conductivity of the control and samples densified at different time.

Treatment Time (min)	Oven-Dry Density (kg/m^3) (n = 3)	Thickness (mm) MC = 0% (n = 4)	Thermal Conductivity (W·m^{-1}·K^{-1}) (MC = 0%) (n = 4)	Thermal Conductivity (W·m^{-1}·K^{-1}) (MC = 6%) (n = 4)	Thermal Conductivity (W·m^{-1}·K^{-1}) (MC = 12%) (n = 4)
0	694.3 (4.1) ^{c,*}	5.43 (0.11)	0.124 (0.008) ^{b,*}	0.130 (0.005) ^{b}	0.140 (0.002) ^{a}
5	848.6 (82.8) ^{b}	4.42 (0.19)	0.097 (0.006) ^{c,d,e,f,g,h}	0.102 (0.005) ^{c,d,e,f}	0.106 (0.006) ^{c}
10	845.4 (76.1) ^{b}	4.08 (0.24)	0.095 (0.009) ^{d,e,f,g,h}	0.098 (0.010) ^{c,d,e,f,g,h}	0.103 (0.009) ^{c,d,e}
15	917.3 (61.0) ^{a,b}	3.78 (0.10)	0.096 (0.009) ^{c,d,e,f,g,h}	0.099 (0.009) ^{c,d,e,f,g}	0.105 (0.008) ^{c,d}
20	948.3 (106.2) ^{a,b}	3.75 (0.16)	0.094 (0.005) ^{e,f,g,h}	0.096 (0.005) ^{c,d,e,f,g,h}	0.101 (0.005) ^{c,d,e,f,g}
25	947.1 (43.5) ^{a,b}	3.62 (0.15)	0.092 (0.008) ^{e,f,g,h}	0.094 (0.007) ^{d,e,f,g,h}	0.099 (0.008) ^{c,d,e,f,g}
30	952.7 (81.9) ^{a,b}	3.60 (0.10)	0.091 (0.004) ^{f,g,h}	0.094 (0.002) ^{d,e,f,g,h}	0.099 (0.002) ^{c,d,e,f,g}
35	1039.6 (60.6) ^{a}	3.63 (0.09)	0.090 (0.002) ^{g,h}	0.093 (0.002) ^{e,f,g,h}	0.099 (0.004) ^{c,d,e,f,g}
40	1031.6 (44.2) ^{a}	3.56 (0.12)	0.095 (0.005) ^{d,e,f,g,h}	0.097 (0.005) ^{c,d,e,f,g,h}	0.100 (0.006) ^{c,d,e,f,g}
45	969.5 (65.6) ^{a,b}	3.44 (0.06)	0.088 (0.007) ^{h}	0.093 (0.004) ^{e,f,g,h}	0.096 (0.005) ^{c,d,e,f,g,h}

Values in parenthesis are standard deviations; * Duncan's test results, average values with the same letter indicate no significant difference at $\alpha = 0.05$.

4. Discussion and Conclusions

THM is a feasible process to increase significantly the density of maple wood in a relatively short time. Most of the oven-dry density increase occurred within the first 5–15 min of treatment. However, the oven-dry density did not show a linear increase with the decrease of thickness. The samples, densified for 10 min, showed a higher density in the core than at the surface, which might have been caused by the large springback after the press opening. The heat distribution across the transverse direction was more homogeneous for samples densified for a longer time with steam. The intrinsic gas permeability of the control samples was 5 to 40 times higher than that of densified wood. This indicated that the void volume of wood reduced notably after the densification treatment.

The thermal conductivity increased by a 0.5–1.5% increase of moisture content for densified sugar maple wood. The thermal conductivity of densified samples was lower than that of the control samples. The densification time had significant effects on oven-dry density and gas permeability. Both densification time and moisture content had significant effects on thermal conductivity, but their interaction effect was not significant.

Author Contributions: Q.F. carried out the experimental tests and prepared the manuscript. A.C. and A.L. contributed to the experimental design and revised the manuscript. All authors read, discussed and approved the final manuscript.

Funding: This research was funded by the Natural Sciences and Research Council of Canada (NSERC) for funding this research under Discovery Grant No. 121954-2012.

Acknowledgments: The authors thank David Lagueux for technical assistance with the densification process.

Conflicts of Interest: The authors declare no conflict of interest.

References

1. Siau, J.F. *Transport Processes in Wood*; Springer: Heidelberg, Germany, 1984; ISBN 13:978-3-642-69215-4.
2. Dai, C.P.; Yu, C.M.; Zhou, X.Y. Heat and mass transfer in wood composite panels during hot pressing. Part II. Modeling void formation and mat permeability. *Wood Fiber Sci.* **2005**, *37*, 242–257.
3. Comstock, G.L. Directional permeability of softwoods. *Wood Fiber Sci.* **2007**, *1*, 283–289.
4. Denisov, O.B.; Anisov, P.P.; Zuban, P.E. Untersuchung der permeabilität von spanvliesen. *Holztechnologie* **1975**, *16*, 10–14.
5. Hata, T.; Kawai, S.; Ebihara, T.; Sasaki, H. Production of particleboards with steam-injection press. Press V. Effects of particle geometry on temperature behaviors in particle mats and on air permeabilities of boards. *Mokuzai Galdcaishi* **1993**, *39*, 161–168.
6. Von Haas, G.; Steffen, A.; Frühwald, A. Permeability of fibre, particle and strand mats for gas. *Holz Roh Werkst.* **1998**, *56*, 386–392.
7. Lihra, T.; Cloutier, A.; Zhang, S.Y. Longitudinal and transverse permeability of balsam fir wetwood and normal heartwood. *Wood Fiber Sci.* **2000**, *32*, 164–178.
8. García, R.A.; Cloutier, A. Characterization of heat and mass transfer in the mat during the hot pressing of MDF panels. *Wood Fiber Sci.* **2005**, *37*, 23–41.
9. Defo, M.; Cloutier, A.; Fortin, Y. Modeling vacuum-contact drying of wood: The water potential approach. *Dry. Technol.* **2000**, *18*, 1737–1778. [CrossRef]
10. Thömen, H. Modeling the Physical Processes in Natural Fiber Composites during Batch and Continuous Pressing. Ph.D. Thesis, Oregon State University, Corvallis, OR, USA, May 2000.
11. Gu, H.M.; Zink-Sharp, A. Geometric model for softwood transverse thermal conductivity. Part I. *Wood Fiber Sci.* **2005**, *37*, 699–711.
12. Simpson, W.; TenWolde, A. Physical properties and moisture relations of wood. In *Wood Handbook: Wood as an Engineering Material*; Ross, R.J., Ed.; USDA Forest Service, Forest Products Laboratory: Madison, WI, USA, 1999; pp. 3–17.
13. Troppová, E.; Švehlík, M.; Tippner, J.; Wimmer, R. Influence of temperature and moisture content on the thermal conductivity of wood-based fibreboards. *Mater. Struct.* **2015**, *48*, 4077–4083. [CrossRef]
14. Griffiths, E.; Kaye, G.W.C. The measurement of thermal conductivity. *Proc. R. Soc. A* **1923**, *104*, 71–98. [CrossRef]
15. MacLean, J.D. Thermal conductivity of wood. *Heat. Pip. Air Cond.* **1941**, *13*, 380–391.
16. Suleiman, B.M.; Larfeldt, J.; Leckner, B.; Gustavsson, M. Thermal conductivity and diffusivity of wood. *Wood Sci. Technol.* **1999**, *33*, 465–473. [CrossRef]
17. Haselein, C.R. Numerical Simulation of Pressing Wood-Fiber Composites. Ph.D. Thesis, Oregon State University, Corvallis, OR, USA, January 1998.
18. De Moura, L.F.; Hernández, R.E. Evaluation of varnish coating performance for two surfacing methods on sugar maple wood. *Wood Fiber Sci.* **2005**, *37*, 355–366.
19. Fang, C.H.; Mariotti, N.; Cloutier, A.; Koubaa, A.; Blanchet, P. Densification of wood veneers by compression combined with heat and steam. *Eur. J. Wood Wood Prod.* **2012**, *70*, 155–163. [CrossRef]

20. Fu, Q.; Cloutier, A.; Laghdir, A. Optimization of the thermo-hygromechanical (THM) process for sugar maple wood densification. *BioResources* **2016**, *11*, 8844–8859. [CrossRef]
21. Siau, J.F. *Wood: Influence of Moisture on Physical Properties*; Department of Wood Science and Forest Products, Virginia Polytechnic Institute and State University: Blacksburg, VA, USA, 1995; ISBN 13:9780962218101-10:0962218103.
22. Fu, Q.; Cloutier, A.; Laghdir, A. Effects of heat and steam on the mechanical properties and dimensional stability of thermo-hygromechanically densified sugar maple wood. *BioResources* **2017**, *12*, 9212–9226. [CrossRef]
23. Finnish ThermoWood Association. *ThermoWood Handbook*; Finnish ThermoWood Association: Helsinki, Finland, 2003.

MDPI
St. Alban-Anlage 66
4052 Basel
Switzerland
Tel. +41 61 683 77 34
Fax +41 61 302 89 18
www.mdpi.com

Fibers Editorial Office
E-mail: fibers@mdpi.com
www.mdpi.com/journal/fibers